Managing Innovation and Standards

"Academics, managers but also policy makers with an interest in managing standards and innovation should read this book. It is unique in providing in-depth insights into all three relevant levels—company, industry, and the innovation's wider context. Based on a detailed study of a best-practice case, the book develops a clear grounded theory and gives useful advice about how intertwined activities on these three levels can lead to aligning innovations, standards, and regulation."

—Knut Blind, *Professor of Innovation Economics, Faculty of Economics and Management, Technical University of Berlin, Germany, and co-editor of the* Handbook of Innovation and Standards

"How exactly standardisation and innovation are related is still far from being fully understood. This book makes an important contribution towards a better understanding of this relation. Most notably, it shows what firms acting in a market that is subject to both regulation and incumbent standards can do to successfully introduce a radical innovation. As such, the book is interesting for both scholars and practitioners."

—Kai Jakobs, *RWTH Aachen University, Founding Editor of the* International Journal of Standardization Research

Paul Moritz Wiegmann

Managing Innovation and Standards

A Case in the European Heating Industry

Paul Moritz Wiegmann
Rotterdam School of Management
Erasmus University Rotterdam
Rotterdam, The Netherlands

ISBN 978-3-030-01531-2   ISBN 978-3-030-01532-9   (eBook)
https://doi.org/10.1007/978-3-030-01532-9

Library of Congress Control Number: 2018956730

This Palgrave Pivot imprint is published by the registered company Springer Nature Switzerland AG
The registered company address is: Gewerbestrasse 11, 6330 Cham, Switzerland

PREFACE

How can innovators manage the seemingly paradoxical relationship between creating radical innovations and complying with external requirements that aim to fix solutions in place? Although businesses face this question whenever they want to bring a new product to market, there is surprisingly little research on the topic. This observation motivated me to investigate how innovative companies deal with standards, as a key example of the external requirements that businesses face.

A review of the literature in Chapter 1 shows that standards indeed have a substantial impact on innovation. Depending on the specific standards, these effects can be positive (e.g. facilitating market access, defining interfaces to supporting infrastructures), but also hinder innovation (e.g. through lock-in). Although the relationship between innovation and standards is not as paradoxical as it first seems, literature confirms its importance for innovators.

To understand how they address this topic, I conducted an in-depth grounded theory case study of the micro Combined Heat and Power (mCHP) technology's development in Europe. As Chapter 2's introduction to the case shows, this radical sustainable innovation is ideal for understanding standards in the context of innovation. Based on in-depth interviews with the key involved actors, I was able to trace in much detail how the technology, standards, and regulation co-evolved.

Studying the case yielded some unexpected insights: It shows that standards' link to regulation can be more central than the literature suggests (Chapter 3). It also suggests that aligning innovations, standards, and regulation is not limited to the company itself. While I observed many company-internal activities on the topic (Chapter 4), interactions between companies and a multitude of other actors are also vital to the case (Chapter 5). Together with the involvement of industry-external actors documented in these chapters, the case shows managing standards and regulation in innovation contexts to be a highly dynamic and potentially contentious process.

These dynamics result from a key property of standards that became apparent in the study: Standards provide certainty and technical detail on (often vaguely defined) requirements from regulation and societal needs. This makes standards (almost) indispensable for innovation, as they create a stable foundation to work on. However, this also means that even standards which focus on seemingly small technical details (e.g. a formula for calculating energy efficiency, see Chapter 5) can cause substantial conflicts between innovators, governments, and other stakeholders.

These insights culminate in a grounded theory (Chapter 6) that answers the question posed at the outset. This theory shows how innovators can position themselves in their industry and its wider context to align innovations with standards and regulation. In doing so, it distinguishes between active and passive approaches to standardisation and regulation. These approaches determine how freely companies can innovate. Chapter 6 also highlights key supporting elements inside the company (e.g. awareness, expertise) and at industry level (e.g. supporting institutions). The grounded theory explains how they contribute to managing standards and regulation in such a way that innovators can introduce their product to the market.

Chapter 7 concludes the book by discussing the findings in light of the literature and giving clear managerial advice to innovative companies and other actors involved in innovations, such as industry associations. While the study started out with a focus on standardisation—as evident from Chapter 1—the unexpected insights make it relevant for broader theories. For example, they highlight standards' importance for socio-technical systems, and underline the need for rules and restrictions for markets' functioning. Chapter 7 also discusses these links and outlines their implications for future research.

I hope that readers find these discoveries as exciting as I do, and enjoy reading this book as much as I did writing it.

Rotterdam, The Netherlands   Paul Moritz Wiegmann
August 2018

ACKNOWLEDGEMENTS

Writing this book concludes an insightful learning process about how companies approach innovation and standards. It would not have been possible without the kind help of a number of people. First, I would like to thank all interviewees for sharing their knowledge about the case so generously with me. Professor Henk de Vries (Rotterdam School of Management, Erasmus University) accompanied me with his critical and detailed expert opinion and knowledge. I am also grateful to Dr. Ursula Lohr-Wiegmann for her support and advice throughout the research process. Last, but not least, Dennis Möller and Nina Laenen were of great help in coding my data. In addition, this study benefited from feedback on earlier versions at the 2015 IEEE-SIIT conference in Sunnyvale, CA; the 2016 EURAM conference in Paris, France; the 2016 EURAS conference in Montpellier, France; the 2017 DRUID conference in New York, NY, and a presentation in April 2018 at Yonsei University, Seoul, Korea. Two anonymous reviewers provided additional helpful feedback. Any remaining errors and omissions are the author's responsibility.

The open access fee was jointly funded by Erasmus University Rotterdam and the Erasmus Open Access Fund.

CONTENTS

LIST OF FIGURES

List of Tables

CHAPTER 1

Introduction: Rooting the Study in the Theoretical Context

Abstract Businesses face regulation, standards, and other external requirements from their operating environments. Taking the example of standards, the chapter reviews findings of these requirements' substantial impacts on innovation and new product development. Depending on the specific standard, these impacts can be positive (e.g. facilitating market access, defining interfaces to supporting infrastructures) or negative (e.g. causing lock-in). This makes standards a key topic for innovators to address. This chapter lays the theoretical foundation for the study by reviewing the limited existing literature on managing standards. Previous company-level studies of standards mostly do not address innovation contexts. Existing industry-level studies on innovation and standards provide few relevant insights for new product development contexts. The chapter concludes by outlining important theoretical gaps that the book addresses.

Keywords Innovation · Effects of standards on innovation · Managing standards in innovation · Standardisation · Impacts of standards

In their operations, businesses face regulation, standards, and other requirements from their operating environments. While some aspire to create free markets with as little external influence as possible (see Friedman, 1962; Krugman, 2007), others argue that such completely free markets are an illusion because they are embedded in societies that

© The Author(s) 2019
P. M. Wiegmann, *Managing Innovation and Standards*,
https://doi.org/10.1007/978-3-030-01532-9_1

impose limitations on actors' behaviour (Fligstein & McAdam, 2012; Polanyi, 2001; Stiglitz, 2001). This implies that such requirements need to be carefully managed to ensure that businesses succeed within these boundaries. In this context, we want to understand how innovative companies manage standards—as an important example of such external requirements—while they are developing new products.

Standards have a profound impact on the development of new technologies, services, and other novel ideas. Extant literature finds that standards are often important factors supporting innovations but can also hinder in other cases. The arguably most fundamental positive effect is that standards often facilitate or even enable innovative products' and services' entry into the market. Other positive effects include, for example, the ability of standards to diffuse knowledge (e.g. Blind & Gauch, 2009; Swann, 2010), standards' potential for facilitating collaboration (e.g. Allen & Sriram, 2000), and their role in creating bandwagons for new technologies (e.g. Belleflamme, 2002; Farrell & Saloner, 1985). On the other hand, examples for standards' negative effects include their potential to restrict creativity and the implementation of new ideas (e.g. Kondo, 2000; Tassey, 2000), as well as the danger that they lock users into using old technologies (e.g. Allen & Sriram, 2000; Tassey, 2000).

These potentially far-reaching effects imply that innovators need to manage standards carefully so that they support, rather than hinder, innovation. Extant literature considers how standards can co-evolve with new technologies to facilitate their emergence (Blind & Gauch, 2009; Featherston, Ho, Brévignon-Dodin, & O'Sullivan, 2016; Ho & O'Sullivan, 2017). These studies focus on the timing when specific types of standards are required to support a technology's further development and on technology roadmapping approaches that can help develop strategies for standardising new technologies. They therefore mostly look at new standards needed for an emerging technology and pay little attention to already existing standards that might affect an innovation and to the processes needed to develop and/or adapt standards for the innovation. This is an important limitation of the extant literature because many of the negative effects of standards found in literature, such as lock-in or limitations for creativity, arise in situations where an innovation is confronted with existing standards. Furthermore, these situations may be particularly challenging to manage because of the dynamics and resistance innovators are likely to encounter when challenging existing standards that may still serve the interests of other actors (see Wiegmann, de Vries, & Blind, 2017).

To generate insights into how companies deal with both existing and new standards, we conduct an exploratory case study of a major innovation within an established industry where many standards apply. In this study, we take the perspective of innovating companies to understand how they manage this topic and its potentially important ramifications for their work. We study the micro Combined Heat and Power (mCHP) technology in the European heating industry. In this case, several companies developed new products in parallel, which were based on the mCHP technology. These products were aimed at existing markets where relevant standards already existed but only partly supported the new technology. Our study shows in detail how this innovation was affected by various standards. Our study also explores how these companies managed the relevant existing and new standards, which industry dynamics resulted from their activities and how these events impacted on the companies' new product development (NPD) activities.

Based on this in-depth study, we develop new theory about managing the co-evolution of innovation with standards and regulation. The resulting theoretical contributions are based on the fundamental finding that activities related to aligning an innovation with relevant standards and regulation occur on three nested levels: (1) the company, which is part of (2) an industry, which in turn is situated in (3) a wider context. Building on this insight, we identify company- and industry-level activities, which are needed to effectively use standards and regulation to align the innovation with needs and demands originating from the wider context. We also pinpoint supporting factors that are needed to carry out these activities successfully and establish through which channels events at each level impact on what happens on the other two levels. We therefore contribute a more detailed and dynamic view to the debate on how to manage standards in innovation contexts, both at company and industry levels.

To firmly root our study in previous findings, we provide a more detailed review of the literature that we summarised in the previous paragraphs. We first look into the extant findings on the links between standards and innovation (Sect. 1.1). Following this discussion, we consider existing insights on how standards can be managed in innovation contexts in Sect. 1.2, which culminates in identifying several important theoretical gaps that motivate the study.

1.1   STANDARDS' EFFECTS ON INNOVATION

Standards, which according to de Vries's (1999, p. 15) definition specify "a limited set of solutions (...) to be used repeatedly", at first sight appear to oppose innovation which aims to create new solutions rather than reuse a limited set of existing ones. In their literature reviews, Dahl Andersen (2013) and Swann and Lambert (2017) found many different ways in which standards impact on innovation. Despite the intuitive expectation that standards are at odds with innovation, Dahl Andersen (2013) reports that around 60% of papers included in his review found a positive link between standards and innovation.

Standards can be distinguished according to their economic functions which include (1) specifying interfaces and providing compatibility; (2) defining minimum quality and safety requirements; (3) reducing variety; (4) disseminating information; and (5) defining measurements (Blind, 2004, 2017; Swann, 2010). Egyedi and Ortt (2017) provide a further refined classification, according to which all standards have the primary functions of (1) reducing variety and (2) providing information. They then identify secondary functions, according to which standards can be distinguished: (1) ensuring compatibility; (2) providing reference measures and defining measurement methods; (3) establishing classifications and (4) codifying behaviour protocols (Egyedi & Ortt, 2017). The impacts of standards differ substantially, depending on which of these categories they fall into (Blind, 2004, 2017; Egyedi & Ortt, 2017; Swann, 2010). Consequently, most of the literature that we cite below focuses on specific types of standards and their effects.

Standards can also be distinguished according to whether they are 'design based' (prescribing a particular specification) or 'performance based' (requiring a certain performance level without specifying how this should be achieved) (Tassey, 2000). Generally speaking, design-based standards are more often constraining for innovation whereas performance-based standards usually are more supporting for innovation (Tassey, 2000). This distinction is therefore similarly important to the distinction between the economic functions for understanding the effects of standards on innovation.

Effects of standards occur at all stages of innovation. They affect the incentives for companies to innovate (e.g. de Vries & Verhagen, 2016; Maxwell, 1998); have implications for the technological development process (e.g. Allen & Sriram, 2000; Blind & Gauch, 2009); and

Table 1.1 Overview of standards' potential effects on innovation

Effects	Positive	Negative
On NPD process	• Providing information • Specifying clear testing and performance guidelines • Facilitating collaboration and division of labour	• Limiting available options for the technology's development • Necessitating collaboration and coordination between actors
On diffusion	• Providing legitimacy and market access • Supporting the emergence of bandwagons and building critical mass • Providing opportunities for generating revenues from the innovation through IPR licensing • Supporting the creation and utilisation of complementary assets and supporting infrastructures	• Preventing or hindering market access • Locking markets into obsolete technologies

Source Author's summary of literature

influence the innovation's eventual diffusion in the market (e.g. Allen & Sriram, 2000; Tassey, 2000). Since our research question concerns the management of standards in the NPD process, i.e. after the decision to innovate has been made, we are particularly interested in the effects of standards on the latter two phases. We provide an overview over these effects in Table 1.1 and outline them in more detail in Sects. 1.1.1 and 1.1.2.

1.1.1 Standards' Effects on the New Product Development Process

Standards play a key role in supporting the development of a new technology. They contribute to the institutional foundations between the involved actors and give them a common understanding of the technology (Bergholz, Weiss, & Lee, 2006; Blind & Gauch, 2009; Foray, 1998; Van de Ven, 1993). More concretely, three key effects of standardisation on NPD activities have been documented in the literature: (1) limiting options available to innovators; (2) acting as a source of information, including about performance requirements; and (3) facilitating (and sometimes requiring) collaboration and division of labour in innovation.

1.1.1.1 Standards Limiting Available Options

The first (and most obvious) effect of standards is limiting the options that are available to an innovation's developers and restricting their choices and freedom in designing their product (e.g. Kondo, 2000; Tassey, 2000). Paradoxically, this may be positive in some situations because it can reduce the search costs involved in solving technological problems (Foray, 1998); ensure that different parties working on an innovation follow a common direction (Swann, 2010); and guide individual actors' investments (Van de Ven, 1993). Furthermore, the degree to which standards limit the available options differs depending on whether they are design- or performance based: While design-based standards are very restrictive, performance-based standards leave more freedom (Kondo, 2000; Tassey, 2000). Process standards that are written in this way may even increase creativity and motivation and thus lead to superior results (Kondo, 2000).

1.1.1.2 Standards as an Information Source

Second, standards are a useful source of information for innovation (Allen & Sriram, 2000; Bergholz et al., 2006; Blind, 2004; Blind & Gauch, 2009; Featherston et al., 2016; Schmidt & Werle, 1998; Swann, 2010; Van de Ven, 1993). This information is particularly important when developing new technologies and/or products in networked industries where the innovation must work seamlessly with other elements of a network (Bergholz et al., 2006; Blind, 2004; Schmidt & Werle, 1998). Standards can also be used to disseminate results from basic research to facilitate their application in an innovation (Allen & Sriram, 2000; Bergholz et al., 2006; Blind & Gauch, 2009) and can facilitate the interface between developing new products and developing the needed production processes to manufacture them at large scale (Lorenz, Raven, & Blind, 2017). This also makes standards a potential external source of innovation for open innovation, in addition to the ones outlined by West and Bogers (2014).

Especially for design-based standards, the degree to which this information is useful for developing innovations depends on two factors. (1) Technological solutions included in standards are sometimes related to someone's intellectual property rights (IPR). If this is the case, this IPR must be available for licensing so that the information can be used by actors who are developing an innovation (Tassey, 2000). (2) The information disseminated through the standard should be up to date and have been included in the standard when the underlying technology was sufficiently mature. Outdated information may no longer be useful and even lock innovators into using old technological solutions (Allen & Sriram, 2000; Swann, 2010; Tassey, 2000). Information included in standards that were passed too early in a technology's lifecycle may constrain its further development or be incomplete (Blind & Gauch, 2009; Tassey, 2000).

When standards are performance based, the information included in them is valuable to innovators because it specifies targets that an innovation has to meet (Abraham & Reed, 2002; de Vries & Verhagen, 2016; Swann, 2010). However, when these requirements and testing procedures are not harmonised internationally, they can also lead to substantial additional efforts. In such cases, required tests need to be repeated for each country where the innovation is intended to be sold (Abraham & Reed, 2002).

1.1.1.3 Standards Facilitating Collaboration and Division of Labour

Third, standards support and sometimes also require collaboration and division of labour in innovation. Standardised interfaces in complex system enable companies to focus their innovations on particular elements of these systems (Chen & Liu, 2005; Tassey, 2000) and base these innovations on complementary assets provided by other parties (see, e.g. Teece, 1986, 2006). Furthermore, standardised interfaces between companies also facilitate collaboration between them in innovation projects, as Allen and Sriram (2000) demonstrate in the case of the Boeing 777's development. However, standards may also necessitate collaboration and a systemic approach to innovation when the requirements set in performance standards are higher than what one actor can achieve individually, as de Vries and Verhagen's (2016) case of the Dutch building sector shows. In such cases, achieving the required performance level may invoke reconfiguring a system's underlying architecture, rather than only innovating parts of it and therefore require the input of all actors who are involved in the system (de Vries & Verhagen, 2016). From an innovator's point of view, this may signify substantial additional cost and effort.

1.1.2   Standards' Effects on Technology Diffusion

In addition to the effects on developing an innovation, standards also may enable or hinder the innovation's eventual success in the market.

While they have the positive effect of providing legitimacy and access to the market and supporting the development of complementary assets, they potentially can also impede an innovation's diffusion by causing lock-in.

1.1.2.1 Standards Providing Legitimacy, Market Access and Supporting Complementary Assets

Standards are central to framing markets for technologies by defining and codifying rules, norms, and values that actors in these markets should follow (Delemarle, 2017). By doing so, they fulfil a key function of legitimising solutions (see Botzem & Dobusch, 2012; Tamm Hallström & Boström, 2010). This legitimation is likely to be particularly important for innovations where actors may be sceptical and still uncertain about the benefits. In such a context, testing the product according to respected standards can help signal an innovation's quality to the market (Tassey, 2000) and thus legitimise it. In Europe, such testing standards can also help to prove an innovation's regulatory compliance to the authorities and therefore provide access to the market. In technological areas that are covered by the 'New Approach', following standards which have been recognised by the European Commission gives actors a 'presumption of conformity' (Borraz, 2007; European Parliament & Council of the European Union, 2002; Frankel & Galland, 2017).

An additional way in which standards can contribute to an innovation's legitimacy is by signalling that it is likely to be adopted by many players (Farrell & Saloner, 1985; Van de Ven, 1993). This expectation is based on the broad support needed for a solution to emerge as a standard (see Wiegmann et al., 2017) but also on other factors, such as the role that standards play in government procurement and the associated demand (Blind, 2008; Edler & Georghiou, 2007; Rosen, Schnaars, & Shani, 1988). Standards can therefore help to "build focus and critical mass in the formative stages of a market" (Swann, 2010, p. 9) , prevent market fragmentation and support exploiting network effects (Bergek, Jacobsson, Carlsson, Lindmark, & Rickne, 2008). If standards contribute to the widespread use of an innovation in this manner, this can also lead to substantial additional revenues for the innovation's developers from licensing fees paid on IPR that is declared standard essential (Kang & Motohashi, 2015).

Finally, innovations often rely on complementary assets and/or supporting infrastructures for their success (Teece, 1986, 2006). In addition to creating critical mass which encourages others to supply these assets (Rosen et al., 1988), standards can also play a more direct role in their provision. By disseminating information about the innovation, standards help others to produce the required complementary assets in the manner outlined in Sect. 1.1.1 (Blind & Gauch, 2009; Schmidt & Werle, 1998). When standards are incorporated into the innovation's development in this manner, they also allow the innovation to make use of existing complementary assets and supporting infrastructures.

1.1.2.2 Standards Causing Lock-In

Although standards can contribute positively to an innovation's diffusion, they can also create lock-in that prevents users from adopting the new product (e.g. Allen & Sriram, 2000; David, 1985; Farrell & Klemperer, 2007; Tassey, 2000). A classic example of lock-in is the QWERTY keyboard which persists in usage despite better alternatives being available (e.g. Allen & Sriram, 2000; David, 1985). In cases of lock-in, large parts of the market use a solution based on an outdated standard and face high switching costs (David, 1985; Rosen et al., 1988). These switching costs prevent the users from adopting the innovation, even if it is superior to the solution prescribed by the existing standard.

1.2 MANAGING STANDARDS IN INNOVATION CONTEXTS

The effects of standards on innovation outlined in Sect. 1.1 make them an important element of innovation management. In Sect. 1.2.1, we summarise the limited available literature about company-level standards management. Other literature provides some insights into how standards and innovation co-evolve on the industry level (see Sect. 1.2.2) but neglects important dynamics, which may, e.g. result from conflicting stakes. In Sect. 1.2.3, we argue why these dynamics are likely to occur and what implications they may have for managing standards in innovation contexts. Finally, we summarise the important gaps in the literature that form the basis for our study (Sect. 1.2.4).

1.2.1 Managing Standards on the Company Level

Although literature about managing standards on the company level mostly does not specifically address innovation (the paper by Großmann,

Filipović, & Lazina, 2016 being a notable exception), several authors (Adolphi, 1997; Axelrod, Mitchell, Thomas, Bennett, & Bruderer, 1995; Blind & Mangelsdorf, 2016; Foukaki, 2017; Jakobs, 2017; van Wessel, 2010; Wakke, Blind, & De Vries, 2015) offer insights that are also likely to apply in this context. On a fundamental level, they argue that managing standards needs to be aligned with the overall business strategy. To do so, companies should formulate a standardisation strategy (Adolphi, 1997; Großmann et al., 2016), which may be driven by the company's organisational culture (Foukaki, 2017). Based on this, organisational structures need to be put in place that enable activities on the tactical and operational levels which help achieve the strategic goals (Adolphi, 1997; Foukaki, 2017). The resulting organisational structures need to facilitate a number of day-to-day tasks, such as applying standards, monitoring the application of standards within the firm, informing company-internal stakeholders about standards, and influencing standard development processes (Adolphi, 1997). In the specific innovation context, Großmann et al. (2016) argue that these day-to-day tasks mainly concern screening existing standards regarding their relevance for the innovation and activities related to feeding the innovation's results into new standard development. These activities should then be related to specific decision points in the NPD process (Großmann et al., 2016).

Adolphi (1997) argues that companies face 'make-or-buy-decisions' whenever they encounter a situation where a standard is needed, meaning that they can either implement existing standards or contribute to developing new ones.[1] Decisions to engage in standard development can be based on a number of strategic motives, such as facilitating market access, influencing regulation, seeking knowledge, maximising compatibility, or enhancing prospects in international trade (Axelrod et al., 1995; Blind & Mangelsdorf, 2016; Foukaki, 2017; Jakobs, 2017; Wakke et al., 2015). Following this decision, companies need not only participate in forums where standards are developed but also carry out supporting activities, such as eliciting requirements and defining success criteria according to which the standardisation work's outcomes can be evaluated (Jakobs, 2017).

---

[1]Adolphi (1997) focuses on company-internal standardisation. Based on this background, he suggests a third option of developing company-internal standards. Due to our study's focus on the effects of (inter)national standards, we do not review this aspect of his work.

Alternatively, companies can implement already-existing standards. Van Wessel (2010) identifies four necessary activities in this context, each of which needs to be carefully managed: (1) selecting appropriate standards, (2) implementing them, (3) using the standard, and (4) assessing the outcomes. One key aspect of managing these activities is that all affected company-internal stakeholders need to be involved throughout the process in order to ensure alignment with their needs (van Wessel, 2010).

1.2.2 Co-evolving Innovation and Standards at Industry Level

Because standards are key to framing markets for new innovations, they need to co-evolve with emerging technologies (Delemarle, 2017). Some existing studies consider how this (should) happen at the industry level (Blind & Gauch, 2009; Featherston et al., 2016; Ho & O'Sullivan, 2017). Blind and Gauch (2009) argue that specific types of standards (e.g. semantic standards or interface standards) are needed at various stages as a technology evolves from pure basic research to its application in the market. In this context, the interface between the R&D process and standardisation and the involvement of scientists and practitioners are particularly important to ensure that standards, reflecting both the state of research and practical applications, are developed (Blind & Gauch, 2009). A technology roadmapping approach can be used to plan such a process and ensure that the necessary standards are developed at the right point in time (Featherston et al., 2016; Ho & O'Sullivan, 2017). Featherston et al. (2016) and Ho and O'Sullivan (2017) develop a framework that links required standards to specific activities in the technological trajectory and allows actors to plan the standardisation process(es) alongside a technology's development.

These existing approaches to co-evolving standards and innovation at industry level focus on the development of new standards needed to support an innovation. While there are cases where scientific discoveries lead to an entirely new technology being developed with no pre-existing standards, such as the example of nanotechnology that Delemarle (2017) and Blind and Gauch (2009) use, many innovations are developed in areas where relevant standards already exist. If these standards have the positive effects on innovation cited in Sect. 1.1, this is not an issue. However, standards with negative effects such as lock-in, need to be updated to increase an innovation's chances of success. In this context, current literature offers some insights into how standards can be changed when needed.

Changes to standards occur on a regular basis—for example, 40% of the standards studied in a study of IT standards were subject to changes at some point in their lifecycles (Egyedi & Heijnen, 2008; Schmidt & Werle, 1998). Such an evolution of standards often follows out of innovations and is driven by four key reasons: (1) new user requirements; (2) anticipation of new technology features; (3) requirements from new technological development, and (4) new applications of existing technologies (Egyedi, 2008). These changes can manifest themselves in deviating ways of implementing the standard (Egyedi & Blind, 2008) which implies that there is no formal process to change the standard and an alternative implementation may become a de facto standard if it is adopted by a large number of players (see, e.g. den Uijl, 2015). Furthermore, these changes can also result from more formalised, and therefore also more manageable, processes. Many standard setting organisations (SSOs) have procedures to update standards, e.g. by releasing updated versions and/or withdrawing outdated standards and replacing them with new documents (Egyedi & Blind, 2008). Due to the time needed for these procedures, these changes in standards are likely to occur with some delay after the corresponding technological development (see Adolphi, 1997, p. 41).

1.2.3   Dynamics Affecting the Management of Standards in Innovation Contexts

Standardisation in innovation contexts often is a contentious issue. The standardisation process is likely to include a range of stakeholders and may also be influenced by external factors, such as societal debates and trends (Delemarle, 2017). When establishing new standards to support an innovation, these actors are likely to attempt influencing standards in a way that gives them an advantage in the innovation's further development (e.g. Blind & Mangelsdorf, 2016; Delemarle, 2017; Rosen et al., 1988; Teece, 2006; Van de Ven, 1993). Furthermore, changing standards frequently leads to issues like added complexity, reduced interoperability, and problems for standard implementation (Egyedi & Heijnen, 2008). Actors with no stake in the innovation may therefore resist changes in standards needed for the innovation's success to avoid such issues.

Such competing interests have strong implications for a standardisation process, e.g. conflicts in SSOs (e.g. Jain, 2012), fierce battles in the market (e.g. den Uijl, 2015), or government involvement in the process (e.g. Meyer, 2012). The resulting dynamics may even be amplified when multiple of the three modes of standardisation (committee based; market based; government based) are involved (Wiegmann et al., 2017). This results in a challenge for innovators to influence standards in such a way that they are eventually supporting, rather than hindering.

1.2.4   Gaps in the Literature

The available literature provides a good foundation for understanding how to manage standards in innovation contexts, but nevertheless leaves important questions unanswered. Our literature review suggests that a more complete understanding is needed of (1) the company level, where the 'managing' is done, and (2) industry-level processes which are likely to result from these management activities but also shape them to some extent. The management of standards in innovation contexts is therefore preferably studied at both levels.

Specifically, we identify three gaps in the literature: (1) The literature on standards management at company level (see Sect. 1.2.1) mostly does not specifically address the context of innovation, even though we show in Sect. 1.1 that this is an area where the impacts of standards on companies' activities are particularly strong. On the other hand, the literature that considers how standards and innovation co-evolve (see Sect. 1.2.2) largely treats companies as 'black boxes' and does not consider the extensive activities that are likely to happen inside them. (2) Given the lack of attention to the company level, the literature on the co-evolution of innovation and standards also misses out on the dynamics within and between the company- and industry levels which we expect to be a major factor in this co-evolution. (3) Finally, the approaches to the co-evolution of standards in innovation contexts cited in Sect. 1.2.2 pay relatively little attention to conflicting interests and the resulting dynamics in the process (see Sect. 1.2.3). Because most innovative products are arguably aimed at existing markets with existing standards, and with actors who may oppose the innovation, such conflicts can be expected to often be critical when managing standards in this context.

These omissions motivate our case study. Our study design, as outlined in Chapter 2, allows us to capture activities on both levels of interest, the resulting dynamics and their effects on an innovation. We therefore contribute a first step towards addressing these three gaps in the literature.

REFERENCES

Abraham, J., & Reed, T. (2002). Progress, innovation and regulatory science in drug-development: The politics of international standard-setting. *Social Studies of Science, 32*(3), 337–369. https://doi.org/10.1177/03063127020 32003001.

Adolphi, H. (1997). *Strategische Konzepte zur Organisation der betrieblichen Standardisierung.* Berlin, Vienna, Zürich: Beuth Verlag.

Allen, R. H., & Sriram, R. D. (2000). The role of standards in innovation. *Technological Forecasting and Social Change, 64*(2–3), 171–181. https://doi.org/10.1016/S0040-1625(99)00104-3.

Axelrod, R., Mitchell, W., Thomas, R. E., Bennett, D. S., & Bruderer, E. (1995). Coalition formation in standard-setting alliances. *Management Science, 41*(9), 1493–1508.

Belleflamme, P. (2002). Coordination on formal vs. de facto standards: A dynamic approach. *European Journal of Political Economy, 18*(1), 153–176. https://doi.org/10.1016/S0176-2680(01)00073-8.

Bergek, A., Jacobsson, S., Carlsson, B., Lindmark, S., & Rickne, A. (2008). Analyzing the functional dynamics of technological innovation systems: A scheme of analysis. *Research Policy, 37*(3), 407–429. https://doi.org/10.1016/j.respol.2007.12.003.

Bergholz, W., Weiss, B., & Lee, C. (2006). *Benefits of standardization in the microelectronics industries and their implications on nanotechnology and other innovative industries.* Retrieved August 14, 2017, from http://www.iec.ch/about/globalreach/academia/pdf/bergholz-1.pdf.

Blind, K. (2004). *The economics of standards—Theory, evidence, policy.* Cheltenham: Edward Elgar.

Blind, K. (2008, September). Driving innovation—Standards and public procurement. *ISO Focus,* pp. 44–45.

Blind, K. (2017). The economic functions of standards in the innovation process. In R. Hawkins, K. Blind, & R. Page (Eds.), *Handbook of innovation and standards* (pp. 38–62). Cheltenham: Edward Elgar. http://doi.org/10.4337/9781783470082.

Blind, K., & Gauch, S. (2009). Research and standardisation in nanotechnology: Evidence from Germany. *The Journal of Technology Transfer, 34*(3), 320–342. http://doi.org/10.1007/s10961-008-9089-8.

Blind, K., & Mangelsdorf, A. (2016). Motives to standardize: Empirical evidence from Germany. *Technovation, 48–49*, 13–24. https://doi.org/10.1016/j.technovation.2016.01.001.

Borraz, O. (2007). Governing standards: The rise of standardization processes in France and in the EU. *Governance, 20*(1), 57–84. https://doi.org/10.1111/j.1468-0491.2007.00344.x.

Botzem, S., & Dobusch, L. (2012). Standardization cycles: A process perspective on the formation and diffusion of transnational standards. *Organization Studies, 33*(5–6), 737–762. https://doi.org/10.1177/0170840612443626.

Chen, K. M., & Liu, R.-J. (2005). Interface strategies in modular product innovation. *Technovation, 25*(7), 771–782. https://doi.org/10.1016/j.technovation.2004.01.013.

Dahl Andersen, F. S. (2013). Standards and innovation: A systematic literature review. In K. Jakobs, H. J. de Vries, A. Ganesh, A. Gulasci, & I. Soetert (Eds.), *EURAS proceedings 2013—Standards: Boosting European competitiveness* (pp. 77–91). Aachen: Wissenschaftsverlag Mainz.

David, P. A. (1985). Clio and the economics of QWERTY. *The American Economic Review, 75*(2), 332–337.

de Vries, H. J. (1999). *Standardization—A business approach to the role of national standardization organizations*. Boston, Dordrecht, and London: Kluwer Academic Publishers.

de Vries, H. J., & Verhagen, W. P. (2016). Impact of changes in regulatory performance standards on innovation: A case of energy performance standards for newly-built houses. *Technovation, 48–49*, 56–68. https://doi.org/10.1016/j.technovation.2016.01.008.

Delemarle, A. (2017). Standardization and market framing: The case of nanotechnology. In R. Hawkins, K. Blind, & R. Page (Eds.), *Handbook of innovation and standards* (pp. 353–373). Cheltenham: Edward Elgar. http://doi.org/10.4337/9781783470082.

den Uijl, S. (2015). *The emergence of de-facto standards*. Rotterdam: Erasmus Research Institute of Management (ERIM). Retrieved from http://hdl.handle.net/1765/1.

Edler, J., & Georghiou, L. (2007). Public procurement and innovation—Resurrecting the demand side. *Research Policy, 36*(7), 949–963. https://doi.org/10.1016/j.respol.2007.03.003.

Egyedi, T. M. (2008). Conclusion. In T. M. Egyedi & K. Blind (Eds.), *The dynamics of standards* (pp. 181–189). Cheltenham: Edward Elgar.

Egyedi, T. M., & Blind, K. (2008). General introduction. In T. M. Egyedi & K. Blind (Eds.), *The dynamics of standards* (pp. 1–14). Cheltenham: Edward Elgar.

Egyedi, T. M., & Heijnen, P. (2008). How stable are IT standards? In T. M. Egyedi & K. Blind (Eds.), *The dynamics of standards* (pp. 137–154). Cheltenham: Edward Elgar.

Egyedi, T. M., & Ortt, J. R. (2017). Towards a functional classification of standards for innovation research. In R. Hawkins, K. Blind, & R. Page (Eds.), *Handbook of innovation and standards* (pp. 105–134). Cheltenham: Edward Elgar. http://doi.org/10.4337/9781783470082.

European Parliament, & Council of the European Union. (2002, January 15). Directive 2001/95/EC of the European Parliament and of the Council on general product safety. *Official Journal of the European Communities, L11,* 4–17. Retrieved from http://eur-lex.europa.eu/LexUriServ/LexUriServ.do?uri=OJ:L:2002:011:0004:0017:en:PDF.

Farrell, J., & Klemperer, P. (2007). Coordination and lock-in: Competition with switching costs and network effects. In M. Armstrong & R. Porter (Eds.), *Handbook of industrial organization* (Vol. 3, pp. 1970–2056). Amsterdam: Elsevier B.V. http://doi.org/10.1016/S1573-448X(06)03031-7.

Farrell, J., & Saloner, G. (1985). Standardization, compatibility, and innovation. *The Rand Journal of Economics, 16*(1), 70–83.

Featherston, C. R., Ho, J.-Y., Brévignon-Dodin, L., & O'Sullivan, E. (2016). Mediating and catalysing innovation: A framework for anticipating the standardisation needs of emerging technologies. *Technovation, 48–49,* 25–40. https://doi.org/10.1016/j.technovation.2015.11.003.

Fligstein, N., & McAdam, D. (2012). *A theory of fields.* New York: Oxford University Press.

Foray, D. (1998). Standards and innovation in technological dynamics. *StandardView, 6*(2), 81–84.

Foukaki, A. (2017). *Corporate standardization management: A case study of the automotive industry.* Lund: Lund University. Retrieved from http://portal.research.lu.se/ws/files/21522119/Corporate_Standardization_Management_A_Case_Study_of_the_Automotive_Industry_Dissertation_2017.pdf.

Frankel, C., & Galland, J.-P. (2017). Markets, standardization and innovation: Reflections on the European Single Market. In R. Hawkins, K. Blind, & R. Page (Eds.), *Handbook of innovation and standards* (pp. 287–301). Cheltenham: Edward Elgar. http://doi.org/10.4337/9781783470082.

Friedman, M. (1962). *Capitalism and freedom.* Chicago: University of Chicago Press.

Großmann, A.-M., Filipović, E., & Lazina, L. (2016). The strategic use of patents and standards for new product development knowledge transfer. *R&D Management, 46*(2), 312–325. https://doi.org/10.1111/radm.12193.

Ho, J., & O'Sullivan, E. (2017). Strategic standardisation of smart systems: A roadmapping process in support of innovation. *Technological*

Forecasting and Social Change, 115, 301–312. https://doi.org/10.1016/j.techfore.2016.04.014.

Jain, S. (2012). Pragmatic agency in technology standards setting: The case of Ethernet. *Research Policy, 41*(9), 1643–1654. https://doi.org/10.1016/j.respol.2012.03.025.

Jakobs, K. (2017). Corporate standardization management and innovation. In R. Hawkins, K. Blind, & R. Page (Eds.), *Handbook of innovation and standards* (pp. 377–397). Cheltenham: Edward Elgar. http://doi.org/10.4337/9781783470082.

Kang, B., & Motohashi, K. (2015). Essential intellectual property rights and inventors' involvement in standardization. *Research Policy, 44*(2), 483–492. https://doi.org/10.1016/j.respol.2014.10.012.

Kondo, Y. (2000). Innovation versus standardization. *The TQM Magazine, 12*(1), 6–10. Retrieved from http://www.emeraldinsight.com/journals.htm?articleid=841925&show=abstract.

Krugman, P. (2007). Who was Milton Friedman? *New York Review of Books, 54*(2), 27. Retrieved from http://givatram.org/bank/content/sikumim/3_2007_08243_08.pdf.

Lorenz, A., Raven, M., & Blind, K. (2017). The role of standardization at the interface of product and process development in biotechnology. *The Journal of Technology Transfer,* 1–37. http://doi.org/10.1007/s10961-017-9644-2.

Maxwell, J. W. (1998). Minimum quality standards as a barrier to innovation. *Economics Letters, 58*(3), 355–360. https://doi.org/10.1016/S0165-1765(97)00293-0.

Meyer, N. (2012). *Public intervention in private rule making: The role of the European Commission in industry standardization.* The London School of Economics and Political Science (LSE). Retrieved from http://etheses.lse.ac.uk/236/.

Polanyi, K. (2001). *The great transformation—The political and economic origins of our time* (2nd Beacon paperback). Boston, MA: Beacon Press.

Rosen, B. N., Schnaars, S. P., & Shani, D. (1988). A comparison of approaches for setting standards for technological products. *Journal of Product Innovation Management, 5*(2), 129–139. https://doi.org/10.1016/0737-6782(88)90004-5.

Schmidt, S. K., & Werle, R. (1998). *Coordinating technology—Studies in the international standardization of telecommunications.* Cambridge, MA: The MIT Press.

Stiglitz, J. E. (2001). Foreword. In K. Polanyi (Ed.), *The great transformation—The political and economic origins of our time* (2nd Beacon paperback, pp. vii–xvii). Boston, MA: Beacon Press.

Swann, G. M. P. (2010). The economics of standardization: An update. Retrieved March 21, 2013, from http://www.bis.gov.uk/assets/biscore/innovation/docs/e/10-1135-economics-of-standardization-update.pdf.

Swann, G. M. P., & Lambert, R. (2017). Standards and innovation: A brief survey of empirical evidence and transmission mechanisms. In R. Hawkins, K. Blind, & R. Page (Eds.), *Handbook of innovation and standards* (pp. 21–37). Cheltenham: Edward Elgar. http://doi.org/10.4337/9781783470082.

Tamm Hallström, K., & Boström, M. (2010). *Transnational multi-stakeholder standardization: Organizing fragile non-state authority*. Cheltenham: Edward Elgar.

Tassey, G. (2000). Standardization in technology-based markets. *Research Policy, 29*(4–5), 587–602. https://doi.org/10.1016/S0048-7333(99)00091-8.

Teece, D. J. (1986). Profiting from technological innovation: Implications for integration, collaboration, licensing and public policy. *Research Policy, 15*(6), 285–305. https://doi.org/10.1016/0048-7333(86)90027-2.

Teece, D. J. (2006). Reflections on "Profiting from Innovation". *Research Policy, 35*(8), 1131–1146. https://doi.org/10.1016/j.respol.2006.09.009.

Van de Ven, A. H. (1993). A community perspective on the emergence of innovations. *Journal of Engineering and Technology Management, 10*(1–2), 23–51. https://doi.org/10.1016/0923-4748(93)90057-P.

van Wessel, R. (2010). *Toward corporate IT standardization management—Frameworks and solutions*. Hershey, PA: Information Science Reference.

Wakke, P., Blind, K., & De Vries, H. J. (2015). Driving factors for service providers to participate in standardization: Insights from the Netherlands. *Industry and Innovation, 22*(4), 299–320. https://doi.org/10.1080/13662716.2015.1049865.

West, J., & Bogers, M. (2014). Leveraging external sources of innovation: A review of research on open innovation. *Journal of Product Innovation Management, 31*(4), 814–831. https://doi.org/10.1111/jpim.12125.

Wiegmann, P. M., de Vries, H. J., & Blind, K. (2017). Multi-mode standardisation: A critical review and a research agenda. *Research Policy, 46*(8), 1370–1386. https://doi.org/10.1016/j.respol.2017.06.002.

CHAPTER 2

Background on Methodology and Case

Abstract The development of micro Combined Heat and Power (mCHP), a radical innovation in the European heating industry, occurred in response to demands for increased energy efficiency and CO_2 emission reductions. This chapter introduces the mCHP case, which provides an excellent understanding of how innovators address standards. The chapter provides an overview over the study's grounded theory approach, which is based on extensive interviews with innovators and other key actors. The chapter also offers important background information about mCHP and the European heating industry. This traditional industry is characterised by its predominantly small- and medium-sized firms and their focus on long-term development.

Keywords Grounded theory · Case study · micro Combined Heat and Power, mCHP · European heating industry · Green technologies

To address the theoretical gaps identified in Sect. 1.2.4, we studied the development of micro Combined Heat and Power (mCHP) technology in the European heating sector. In this chapter, we provide some background information that is helpful for understanding our findings. Section 2.1 outlines our grounded-theory-based methodological approach. Section 2.2 introduces mCHP technology and the setting in which it was developed.

P. M. Wiegmann, *Managing Innovation and Standards*,
https://doi.org/10.1007/978-3-030-01532-9_2

2.1   Grounded Theory Methodology

As outlined in Chapter 1, we are interested in a detailed exploration of how innovators manage external requirements (imposed by standards), the dynamics that result from this, and how this affects NPD activities. Specifically, we want to explore how this occurs on the company- and industry levels and how these two levels interact. The lack of literature addressing these questions makes an in-depth exploratory case study, which uses inductive reasoning to derive a grounded theory, the most suitable research design (Eisenhardt, 1989; Glaser & Strauss, 1973; Yin, 2009). This grounded theory approach allows us to conceptualise patterns that we find across the data to generate our theoretical contribution (Glaser & Strauss, 1973). In Sect. 2.1.1, we explain our case selection. Section 2.1.2 shows how we collected our data. Finally, Sect. 2.1.3 summarises our approach to analysing these data.

2.1.1   Case Selection: Theoretical Sampling

Following Eisenhardt (1989) and Yin (2009), we selected our case on theoretical grounds rather than through random sampling. Following on from our research question and the identified gaps in the literature, we defined five criteria that the case would have to meet. (1) It needed to be a case of an innovation for which both existing standards are relevant and new standards are required. (2) This innovation needed to represent a substantial technological leap. This maximised our chances of observing standards having a major impact on the innovation, and the involved actors' approaches to managing these impacts. (3) Our specific interest in NPD activities also means that the innovation in our case needed to be at a stage when companies developed products intended to be sold on a large scale. The initial fundamental research considered by Blind and Gauch (2009) should therefore already have been concluded. (4) Furthermore, NPD activities concerning the innovation should preferably be pursued in parallel by several companies as this would allow us to compare their potentially different approaches to managing the relevant standards. (5) Finally, for practical reasons, data about the case needed to be accessible and the case should be relatively recent to ensure that informants would be able to recall the needed information.

We found a suitable case which meets all five requirements in the development of micro Combined Heat and Power (mCHP) technology.

Several companies in the European heating industry simultaneously developed innovative natural gas powered central heating boilers, which convert excess heat into electricity, making them embedded units in the case (see Yin, 2009). Standards were relevant, both because interfaces with other supporting infrastructures (e.g. the electrical installation in a building and the electricity grid) are needed for the innovation to be of value and also because important safety and efficiency issues make this a technology that is covered by the European Commission's 'New Approach'.[1] When mCHP was developed, generating electricity was an entirely new feature for the industry, meaning that it was a substantial departure from existing technologies. Nevertheless, there already were several existing standards affecting the technology, because the market that it was aimed at and the supporting infrastructures (gas, electricity, water) were already in place. Lastly, the case also satisfies the practical requirements outlined above.

2.1.2 Data Collection

The largest share of our data was collected in interviews. Following two interviews with existing contacts, we used snowball sampling and contacted actors who we identified as relevant in desk research (e.g. additional companies with mCHP products) and when attending an industry conference. This approach resulted in approximately 26 hours of interviews conducted between April 2015 and August 2017 as detailed in Table 2.1. These interviews gave us insights into the perspectives of all groups of actors who were involved in developing mCHP-related products and/or managing standards to facilitate the technology, as well as perspectives from different countries which are key markets for the new technology.

In order to ensure that the main topics of interest were covered in each interview while leaving the interviewees enough leeway to 'tell their stories', we used a semi-structured format. Gioia et al. (2013) highlight the importance of the interview guideline to ensure that this results in useful data for deriving theoretical patterns. This guideline was adjusted

---

[1] Under the 'New Approach', regulation provides 'essential requirements' for products to be sold on the European market and standards are used to specify these requirements and test methods to assess compliance in detail for specific product groups. Also see our more detailed explanation on this topic and its relevance for the case in Sect. 3.2.1.

Table 2.1 List of interviews in chronological order

Interview No.	Organisation	Interviewee(s)
1	Netherlands-based manufacturer of heating systems, approx. 6500 employees and €1.7bn revenue	Technical innovation manager, responsible for all mCHP-related NPD and standardisation activities
2	Association of the European Heating Industry (EHI)	Technical affairs director, responsible for all mCHP-related activities at the association[a]
3	Engineering research institute at a German university	Researcher, involved in mCHP-related contract research and participating in mCHP-related standardisation committees
4	Germany-based manufacturer of heating systems, approx. 12,000 employees and €2.25bn revenue	Manager responsible for coordinating the company's standardisation activities, involved in mCHP's technological development in a previous role
		Manager responsible for the company's participation in associations on a strategic level, previously head of new technology development
5	Germany-based manufacturer of heating systems, business unit of a conglomerate with approx. 390,000 employees and €73bn revenue	Manager in charge of the business unit's external affairs, relationships with political actors and governments, and work in industry associations
6	Japan-based supplier of fuel cells for mCHP systems, business unit of a conglomerate with approx. 258,000 employees and €57bn revenue	Manager responsible for advising Japan-based R&D department about European standards and representing the company in European standardisation
7	Germany-based self-employed engineering consultant specialised in mCHP, consulting industry actors on the technical implementation of requirements arising from regulation/standards and active in mCHP-related standardisation committees	
8	Netherlands-based certification body, conducting assessment of mCHP devices' conformity to legal requirements	Head of several testing laboratories, including the one conducting mCHP-related conformity assessment
9	Small UK-based supplier of Stirling engines for mCHP systems	Engineer, involved in the company's mCHP-related R&D in various roles since 2000

(continued)

Table 2.1 (continued)

Interview No.	Organisation	Interviewee(s)
10	Germany-based manufacturer of heating systems, approx. 12,000 employees and €2.4bn revenue	Head of technology development for mCHP, involved in mCHP-related R&D at the company since 1997
		Project leader for CHP applications, responsible for regulatory approval of mCHP devices in an earlier role, involved in mCHP-related R&D at the company since 2000
		Project leader, involved in mCHP-related R&D at the company since 1999
11	Germany-based manufacturer of mCHP systems, approx. 30 employees	Managing director
12	UK-based supplier of fuel cells for mCHP systems, approx. 100 employees	Head engineer overseeing all engineering activities at the company
13	Small Switzerland/Italy-based manufacturer of mCHP systems	Manager responsible for regulatory approval of mCHP devices
14	See Interview 1	See Interview 1

[a] A representative of a manufacturer of heating systems was also present during a short part of this interview and participated in the conversation. This person was then interviewed individually during Interview 5

for each interview to cover all important topics (interviewee's involvement in the case, views on relevant standards, companies' processes for managing the topic, interactions with other stakeholders, results of their activities, etc.). Using these guidelines, we obtained detailed accounts of the interviewees' activities in the case and their views on the events.

Where possible, we recorded the interviews and transcribed them verbatim in the language in which the interview was conducted (English for Interviews 1, 8, 9, 12, and 14; German for all other interviews). In addition, some interviewees provided us with internal company documents. Furthermore, we considered European Union policy documents related to the standards in the case which provided us with additional information on the evolution of standards in relation to the European directives that they were supposed to support. A final source of information was attending an industry conference hosted by the European industry association for co-generation of heat and power (COGEN Europe) in March 2016. At this conference, we gained further insights into the major topics of interest for industry actors and gained background information on how mCHP fits into the wider industry context. The conference also provided us with an opportunity to have informal discussions with important actors in the case.

2.1.3   Data Analysis

In line with our study's inductive reasoning, we based our data analysis on a grounded theory approach (Glaser & Strauss, 1973). We initiated our data analysis in parallel to data collection so that the information from earlier interviews could inform subsequent data collection efforts. In order to come closer to Glaser and Strauss's (1973) ideal of developing grounded theory without preconceived notions of existing theory, two assistants performed most of the open coding (see Alvesson & Sköldberg, 2009; Gioia et al., 2013) under the author's supervision. All coding was performed on transcripts in the languages in which the interviews were conducted (German and English, see Sect. 2.1.2) in order to stay as close as possible to the empirical evidence at this stage.

Simultaneously to coding, we started the further data analysis by 'integrating categories', as suggested by Glaser and Strauss (1973, pp. 108–109). Clear themes that later became the key concepts of our theory emerged from the data at this stage, although we did not follow the strict template provided by Gioa et al. (2013). These theoretically saturated (see Glaser & Strauss, 1973, pp. 111–113) key themes are based on the main discussion topics across our interviews and reflect the elements that our interviewees emphasised. Chapters 3, 4, and 5 are structured along these themes and use extensive quotes from the interviews and—where available—supporting evidence from other sources to ensure that our constructs are deeply rooted in empirical observations.[2]

In parallel to identifying these key concepts, we also looked for relationships between them (see Alvesson & Sköldberg, 2009, pp. 68–69; Glaser & Strauss, 1973, pp. 109–113). As suggested by Glaser and Strauss's (1973) description of the constant comparative method, we did so by alternating between noting down our ideas about such links and verifying in the data whether these ideas were supported by the evidence. This verification was based on whether we could identify a plausible explanation for each relationship in the data, for example by comparing different firms (embedded units) in our case, or by searching for interviewees' explanations of the reasons behind certain activities and events. This process ultimately resulted in the theory that we present in Chapter 6 and makes this theory firmly rooted in the empirical observations from our case.

2.2   INTRODUCING THE MICRO COMBINED HEAT AND POWER (MCHP) CASE

As outlined in Sect. 2.1.1, the development of micro Combined Heat and Power (mCHP) is an excellent case to study the management of standards during the development of a new technology. Combined heat and power (CHP) solutions have been developed for all scales, ranging from domestic family homes to large industrial applications. Our case study traces the development of micro CHP (mCHP) which includes all CHP appliances with up to 5 kW electrical output (EHI, 2014). These appliances would typically be used in single-family houses.

The technology is a major innovation in the European heating sector. In addition to providing hot water and heat for buildings, mCHP boilers also generate electricity. This additional functionality represented a major technological leap for the European heating industry which did

---

[2]Where we quote interviews that were conducted in German we translated them at this stage, labelling each translated quote as such.

previously not make any electricity-generating products. In order to provide context for our analysis of how products using this technology were developed and standards were managed during this process, we cover background information that is important for a good understanding of the case. We first portray the European heating industry and mCHP's role for it (Sect. 2.2.1). Following this, we give a brief overview over different technological approaches to mCHP and how the relevance of standards differed for them (Sect. 2.2.2).

2.2.1   The European Heating Industry and the Market for mCHP

Heating of buildings is estimated to be responsible for around 40% of the EU's energy consumption and 36% of its CO_2 emissions (European Commission, 2017). Consequently, boiler manufacturers and other actors in the European heating industry have been facing expectations from the market and political actors to make their products more energy efficient and contribute to efforts to combat climate change. In response to these demands, the European heating industry developed several technologies to eventually succeed the established condensing boilers for domestic applications, including heat pumps, solar thermal systems, and mCHP. Which of these technologies is most energy efficient depends, e.g. on heat demand and the local electric power generation mix where an appliance is installed. The technologies therefore address different market segments. A key advantage of mCHP products compared to heat pumps and solar thermal systems is that they can be integrated in existing buildings more easily if designed in such a way that they match existing infrastructure in buildings. This made mCHP a potentially promising technology to attain higher energy efficiency in the replacement market, which one interviewee described as existentially important for the companies in the industry:

> We live off the existing [building] stock and replacement. The relation between newly built buildings and existing buildings in Germany in a year is approximately 1:10. This means that, for every boiler or heating appliance that we sell into a newly built house, we sell ten into existing buildings. (translated from German)

The European heating industry is distinctive in that the established players and market leaders are mostly owned by the founding families or by foundations with a mission to ensure the business's long-term viability. This gives the companies and the entire industry a long-term outlook which also manifested itself in the way standards were managed during the development of mCHP. However, it also means that the industry is relatively conservative and *"not really known for being particularly innovative [and consisting of] rather traditionally shaped enterprises"* (translated from German).

Developing mCHP brought the involved actors into contact with several new key technological fields (see Sect. 2.2.2) and the players involved in these areas, requiring the industry to adopt new approaches to innovation and standardisation and become more open to dealing with actors outside the industry as outlined in Chapters 4 and 5. Within the industry, these developments were driven by a range of actors. In addition to the boiler manufacturers (OEMs) who developed and eventually sold complete mCHP appliances, suppliers of key components; certification bodies; engineering consultants; industry associations; and research institutes all were involved in the process. The OEMs developing mCHP and the component suppliers included established players in the industry and new entrants which were specifically founded as start-ups to develop mCHP appliances and components. Our interviews cover all key players in the case as well as some more peripheral actors (see the characterisations of companies covered by our interviews in Table 2.1).

2.2.2   Technological Solutions for mCHP

Four technological approaches exist to realise the functionality of mCHP appliances: (1) Stirling engines; (2) fuel cells; (3) internal combustion engines and (4) steam expansion engines (EHI, 2014). While internal combustion engines and steam expansion engines have been barely used for mCHP applications, both products based on Stirling engines and on fuel cells have been developed and marketed.

All interviewed OEMs have been developing fuel-cell-based mCHP appliances, although not all of them have brought them to the market yet at the time of writing. Some OEMs have been developing and offering Stirling-based mCHP appliances in addition. The OEMs that never developed the Stirling technology or exited its development cited technological challenges and doubts about whether mCHP appliances using Stirling engines could reach the same levels of efficiency as those using fuel cells as the reasons behind the decision to only pursue fuel cells. On the other hand, the companies that still have been pursuing the Stirling engine in parallel to fuel cells see the two technologies as catering for distinctive market segments:

> I expect there will be different technologies in parallel, and they could serve different markets segments. That has to do with the question how the ratio is between heat demand and power demand. That's one issue. And especially when the heat demand is high compared to the power demand then nowadays already Stirling engine could be a better solution than the fuel cell.

Technologically, the two approaches are fundamentally different: (1) Appliances with a Stirling engine add this engine (and some control electronics) to a conventional condensing boiler. Such a boiler produces more heat than is needed to cover the demand for heating and hot water. The excess heat is then converted to AC electricity by the Stirling engine which is tuned to the frequency of the national electricity grid (50 Hz in Europe), meaning that the produced electricity can be fed directly into the grid. (2) Fuel-cell-based appliances contain a reformer that extracts hydrogen from natural gas. This hydrogen is then used to power a fuel cell which produces both heat and DC electricity. An inverter converts this DC electricity to AC electricity that can be fed into the electricity grid. In addition, fuel cell appliances usually include a conventional gas boiler to cover peak heat demand.

Some aspects of these technologies were already known to the involved companies and have been used in their products for decades. Particularly, the condensing boiler units that provide the heat for Stirling engines to operate were very similar to the ones used in the industry's existing products. However, both Stirling engines and in particular fuel cells were new and very complex technologies for all actors in the heating industry. Furthermore, regardless of the technological approach to mCHP, its implementation required the industry to get involved in entirely new technological aspects, such as access to the electricity grid, technologies for communication with other devices, or grid stability. These fields presented a steep learning curve, in terms of both technology development and standardisation, as Chapters 4 and 5 show.

Most relevant standards and regulatory requirements (see Chapter 3) applied equally to Stirling- and fuel-cell-based mCHP appliances and had similar implications for both technologies' development. The standards for connecting appliances to the national electricity grid are a key exception to this. Some changes to them that occurred while mCHP was being developed posed additional challenges for devices using Stirling engines but had a smaller impact on the development of fuel-cell-based mCHP (see Chapters 3 and Sect. 5.2 for details).

REFERENCES

Alvesson, M., & Sköldberg, K. (2009). *Reflexive methodology—New vistas for qualitative research* (2nd ed.). Thousand Oaks, CA: Sage.

Blind, K., & Gauch, S. (2009). Research and standardisation in nanotechnology: Evidence from Germany. *The Journal of Technology Transfer, 34*(3), 320–342. https://doi.org/10.1007/s10961-008-9089-8.

EHI. (2014). *Combined heat and power.* Retrieved September 4, 2017, from http://www.ehi.eu/article/combined-heat-and-power.

Eisenhardt, K. M. (1989). Building theories from case study research. *Academy of Management Review, 14*(4), 532–550. https://doi.org/10.5465/AMR.1989.4308385.

European Commission. (2017). *Energy efficiency—Buildings.* Retrieved September 11, 2017, from http://ec.europa.eu/energy/en/topics/energy-efficiency/buildings.

Gioia, D. A., Corley, K. G., & Hamilton, A. L. (2013). Seeking qualitative rigor in inductive research: Notes on the Gioia methodology. *Organizational Research Methods, 16*(1), 15–31. https://doi.org/10.1177/1094428112452151.

Glaser, B. G., & Strauss, A. L. (1973). *The discovery of grounded theory: Strategies for qualitative research* (5th ed.). Chicago, IL: Aldine Publishing Company.

Yin, R. K. (2009). *Case study research—Design and methods* (4th ed.). Thousand Oaks, CA: Sage.

CHAPTER 3

Standards, Regulation and Conformity Assessment for mCHP

Abstract micro Combined Heat and Power (mCHP) relies on standards in around a dozen technical areas, related to topics like product safety, electricity grid access, and environmental performance. This chapter provides an overview over relevant standards and their effect on mCHP. Under the European 'New Approach', many of these standards define 'essential requirements' in line with European regulation. This link makes standards important elements for conformity assessment and proving mCHP appliances' regulatory compliance. Standards are therefore key enablers for mCHP's developers to place the technology on the European market. The chapter concludes with an overview over the effects of standards and regulation on innovation in the mCHP case.

Keywords Standards · European regulation · European new approach Effects of standards on innovation · Conformity assessment Regulatory compliance

Standards, together with regulation and conformity assessment, have been crucial for the development of mCHP. While our study was initially focussing on the role and management of standards for the innovation, it soon transpired from our interviews that they are inextricably linked to European and national regulation and conformity assessment of mCHP appliances. In Sect. 3.1, we outline which standards have been relevant for the technology's development. Section 3.2 explores the link between

© The Author(s) 2019   33
P. M. Wiegmann, *Managing Innovation and Standards*,
https://doi.org/10.1007/978-3-030-01532-9_3

standards and regulation and its effects on mCHP. Following this, we discuss the need for conformity assessment and the role that standards and regulation play in this context (Sect. 3.3). Finally, we shed light on additional effects that standards had on the development of mCHP in Sect. 3.4.

3.1   RELEVANT STANDARDS FOR mCHP

Standards posed requirements for key aspects of mCHP technology, such as product safety, energy efficiency, and connections to the electricity grid, which needed to be fulfilled in order to provide the intended value for buyers and gain approval for market entry. A list of all relevant standards, that were mentioned during the interviews, can be found in Table 3.1. Many of these standards are interrelated.

The standards identified in Table 3.1 broadly fulfilled two main functions for mCHP's development process: The first function is defining the interfaces to link mCHP to complementary technologies, such as the national electricity grid and electrical and gas installations in buildings. These infrastructures were essential to enable the innovation to deliver the new aspects of its value proposition—generating electricity that can be used by a device's owner and/or fed into the electricity grid.

The second main function of standards for the innovation is related to support proving the compliance of mCHP appliances and their components with regulatory requirements (e.g. gas and electrical safety, energy efficiency and requirements for connecting devices to the electricity grid). This function has been key for the development of mCHP, based on the link between standards and regulation in the case, which we outline in detail in Sects. 3.2 and 3.3.

All interviewees stressed the particular importance of the product standard (EN 50465—"Gas appliances—Combined heat and power appliance of nominal heat input inferior or equal to 70 kW") for the development of mCHP. This product standard addresses key elements of the technology, such as safety and energy efficiency, and defines minimum performance requirements for these dimensions of mCHP appliances. It has been key in outlining how mCHP appliances can meet regulatory requirements (see Sect. 3.2) and in supporting the conformity assessment of the appliances (see Sect. 3.3). When the technology's development started, this standard did not exist yet in its current form and did not cover all technological approaches to mCHP:

Table 3.1 Relevant standards for mCHP

Level of the standard	Technical aspects covered	Standard(s)
Links between mCHP appliance and other systems	Connection to the electricity grid	EN 50438[a]; standards developed by ENTSO-E (European Network of Transmission System Operators for Electricity), national grid codes
	Connection to a building's internal electrical wiring	National standards for electrical installations (e.g. VDE-AR-N 4105 in Germany, NEN 1010 in the Netherlands)
	Communication between distributed electricity producing devices (e.g. other mCHP appliances, solar panels) to ensure grid stability	IEC 61850-7-420, VHPready consortium standard
	Quality and composition of natural gas used to operate mCHP appliances	EN 16726
mCHP appliance as a whole	Product standards: cover product safety; energy efficiency; minimum performance requirements	EN 50465 (used for certification of appliances against European regulatory requirements), IEC 62282-3-400, at early stages of the development also DVGW VP 109 and VP 119
	Product safety	IEC 62282-3-100
	Electrical safety	EN 60335
	Quality management standards needed to make the appliance eligible for financial support offered by some national governments	Microgeneration Certification Standard (MCS 2011)
	Standards describing test methods to be used when assessing the product's conformity to regulatory requirements	EN 437 and others

(continued)

Table 3.1   (continued)

Level of the standard	Technical aspects covered	Standard(s)
Components of mCHP appliance[b]	Burners and burner controls	EN 298, EN 13611
	Electrical safety of components	EN 60730
	Product standards for various components, such as gas valves, pressure controllers, shut-off valves, pressure sensors, etc.	

[a]The abbreviation 'EN' stands for 'Europäische Norm' and refers to European standards developed by the European Standardisation Organisations (ESOs) CEN, CENELEC and/or ETSI
[b]The overview over relevant standards for components is incomplete since the product standard EN 50465 refers to 65 other standards on this level, which were not all named individually in the interviews. Nevertheless, this overview gives a good impression of the range of such standards

At first you have to deal with the product standard. But at the moment that we did the development, it wasn't there. We did the development, the basic development, we started by the end of 2005 and at that moment there was no standard.

This initial absence of the key standard had important implications for the technology's development and made writing this standard a priority for the industry in managing the standards related to the innovation, as we outline in Sect. 5.2.2.

3.2   REGULATION FOR MCHP AND ITS RELATIONSHIP WITH STANDARDS

Relevant regulation for mCHP covers the areas of product safety, energy efficiency and grid connections (see Table 3.2 for a list of all regulatory texts that were mentioned as relevant during the interviews). This regulation defines 'essential requirements' which mCHP appliances must meet if they are sold on the European market. In line with the European 'New Approach', these essential requirements laid down in the regulation are formulated on a relatively abstract level and do not prescribe technical details or solutions that need to be implemented to fulfil them. Standards provide important guidance regarding how to reach these requirements, as outlined below.

Table 3.2   Relevant regulation for mCHP

Type of regulation	Regulation	Relevance for mCHP
European Directives	Energy-Related Products Directive (ErP, also referred to as Ecodesign Directive)	Imposes minimum requirements for energy efficiency of energy-using products
	Energy Labelling Directive	Defines a labelling scheme for energy-related products
	Cogeneration Directive (CHP), replaced in 2012 by the Energy Efficiency Directive	Defines measures to increase the EU's energy efficiency targets, including promoting more energy efficient heating systems
	Gas Appliances Directive (GAD)	Imposes requirements for the safety of gas-powered products
	Low Voltage Directive (LVD)	Imposes requirements for electrical safety
	Electromagnetic Compatibility Directive (EMC)	Imposes requirements regarding emitting and accepting electromagnetic interference
	Machinery Directive (MD)	Imposes safety requirements for machines with moving parts (mainly relevant for Stirling-based mCHP appliances)
	European Network Code on Requirements for Generators (NC RfG),	Imposes requirements for connecting to the electricity grid
National regulation	National electricity laws	Define (financial) conditions under which electricity can be fed into electricity grids

3.2.1   Harmonised Standards Providing 'Presumption of Conformity'

Under the European 'New Approach', the high-level requirements formulated in directives are supported by harmonised standards. These standards provide detailed specifications of the essential requirements, such as test methods to be used in assessing whether a product meets the essential requirements. Such harmonised standards are developed by the ESOs following requests by the European Commission. The European Commission then carries out an assessment whether the contents of these standards satisfy the essential requirements. If a standard passes this assessment, it is listed in the Official Journal of the European Union along with the directive against which it is harmonised.

Once a standard has been harmonised in this procedure, complying with the standard gives a product 'presumption of conformity' with the associated European Directives. This means that any product which implements a harmonised standard is assumed to meet the essential requirements imposed by the directive:

> Someone who develops such a product (...) can work with the standards and can then assume that he also fulfils the requirements from the directives in this way. (...) This is called 'presumption of conformity' if a standard is listed under a directive in the Official Journal (...) which helps from a technical point of view. (translated from German)

3.2.2   Fulfilling 'Essential Requirements' Without Relying on Harmonised Standards

Although relying on harmonised standards is a straightforward and commonly used way of proving compliance with regulatory requirements, their use remains voluntary (European Commission, 2017). Manufacturers are also permitted to demonstrate in other ways that they reach a performance level that satisfies the regulation's essential requirements.

A first way of doing so is implementing other standards developed by the European Standardisation Organisations (ESOs—CEN, CENELEC, and ETSI), even if they are not harmonised. These standards are assumed to reflect the current state of technological development, meaning that implementing them in an innovation is seen as following good practice. This also applies to the key product standard in the mCHP case (EN 50465). Due to conflicts between the European Commission and the European heating industry regarding the calculation methods for mCHP appliances' energy efficiency (see Sect. 5.2.2), this standard has not been harmonised yet at the time of writing. Nevertheless, it has emerged as the generally accepted standard detailing the essential requirements from the relevant European Directives for mCHP appliances.

In addition to or as an alternative to relying on standards, manufacturers may also demonstrate their product's equivalent performance to the level described in the standard without using any standard:

> If his [a manufacturer's] product has a solution that is not covered by the standard (...), this is not forbidden. (...) [But] it has to be written in the development documentation that he (...) fulfils the requirements of the directive. (...) When he, as a manufacturer, prints the CE-mark[1] on the device he confirms at this time that all relevant directives are fulfilled (...) and this has been proven through the standard and (...) his own specifications. (translated from German)

Such an approach of not relying on the standard then shifts the burden of proof that the mCHP appliance meets the regulatory requirements to the manufacturer:

> The burden of proof that this [the product fulfilling the essential requirements] is actually the case then lies with him [the manufacturer]. (...) When he uses a harmonised standard, the presumption of conformity applies. This means that if he uses the standard, he may assume that he fulfils the essential requirement. If this [fulfilling the essential requirement] is not the case, the burden of proof does then not lie with him but with the European Commission. This is all about who is liable. (translated from German)

In addition to the issues surrounding liability when deviating from the solutions defined in a standard, taking such an approach would also require substantial additional effort and slow down the NPD process:

> [Standards] rather lead to speeding up a development process, because the requirements are clear. Imagine there were no standards and we only had the directives. Because directives are laws and safety-related laws always exist. (...) Then you first would have to translate: What does such a legal requirement mean for materials, for testing, for technology, for time response? So standards, because they are general specifications, are actually accelerating means for the development. (translated from German)

In practice, the interviewed manufacturers therefore based the designs of their mCHP appliances on standards wherever possible and avoided using other technical solutions which would have required them to

---

[1] By placing the CE-mark on a product, a manufacturer confirms that it meets all European regulatory requirements and has passed all relevant conformity assessment procedures.

demonstrate compliance to regulatory requirements in other ways. This further underlines the importance of standards for the innovation and also had implications for the management of standards, where the industry sometimes invested substantial resources in order to influence standards, rather than implementing alternative solutions into their products (see Chapter 4).

3.3   Assessing Conformity to Essential Requirements in the mCHP Case

Because the essential requirements in the relevant regulation are mandatory (see Sect. 3.2), mCHP appliances can only be sold in the European market once their compliance to these requirements has been proven. While a declaration by the manufacturer, confirming that the requirements are met, is sufficient for many product groups, this is not the case for mCHP. Due to the inherent safety risks of gas-powered appliances, conformity assessment must be carried out by an accredited certification body which has been authorised by the government to carry out this assessment for the relevant European Directives. This party issues a certificate if the requirements are met[2]:

> [For] a gas appliance, a manufacturer cannot simply develop an appliance, produce it, and sell it. He needs third-party certification. This means he must go to an accredited testing laboratory. The product is tested on its conformity, strictly speaking to the directive but in practice to the standard. Then, a notified body issues the certificate. Only once he has this, he can sell it in Europe. (translated from German)

Such independent test laboratories (often referred to by the legal term 'notified bodies') assess the technology and against essential requirements in the relevant directives. Notified bodies choose an appropriate basis for certification which defines both the requirements that mCHP appliances must fulfil and the methods, which are used to assess the fulfilment. Usually, the product standard (EN 50465 in the mCHP case) is

---

[2]In addition, manufacturers can choose to obtain batch approval for their appliances. This means that they are tested according to less strict criteria and allows manufacturers to sell a limited number of appliances before obtaining full certification.

used for this purpose. It defines both requirements and test methods but (at least in theory) test laboratories may also deviate from this:

> This inspector, who is employed by this institute, decides which basis he brings forward or draws upon to conduct the assessment. And in this, he is relatively free. So, if he says... He could still say today 'the 50465 is not sufficient for me'. This would not correspond to the facts, but he could always draw on another standard if this was necessary in his opinion. (translated from German)

This discretion in choosing the basis for the certification process led to different approaches among testing institutes in the early stages of mCHP's development, when EN 50465 did not yet exist in its current form and therefore no standard detailed the essential requirements for mCHP appliances. In interviews with OEMs, we were told about various related standards (e.g. for conventional condensing boilers) being used as a preliminary basis for testing by the notified bodies. Another approach, which was described in an interview with a notified body, was developing a test regime directly based on the relevant directives:

> When we started this process, typically for fuel cell systems, there was no standard. So we had to certify directly on the directive. We have the essential requirements of the directive. So what we did, we created our test plan and said 'okay if you meet this, then we can certify against the Gas Appliance Directive'. So there was a lot of freedom for us, but in the end, as a competent notified body, we had to make a decision 'it's safe enough'. So, we could handle different technologies which were not addressed by standards. But it also means a very good relation between us and the manufacturer to really understand the technology and for them to understand what our safety requirements are.

3.3.1   Standards Providing Certainty for Conformity Assessment

The potentially different approaches to certifying mCHP appliances that could be followed in the absence of standards meant some uncertainty for the NPD process because the exact requirements for market access only became clear when the notified bodies were invoked into the companies' NPD activities. As Sect. 4.2.2 shows, the stages of development at which notified bodies were involved varied between companies, meaning that the magnitude of the resulting uncertainty also differed across actors in the industry. Nevertheless, having standards (in particular EN 50465) in place to provide more detailed information about essential requirements, as outlined in Sect. 3.2, helped all involved parties' NPD activities because this reduced leeway for different interpretations of the essential requirements:

> It is very important for industry that not everybody interprets the directive differently every day and at the end the certification laboratory differently than the manufacturer. (translated from German)

In this way, standards provided important information about required performance and test procedures to prove this performance which could be used during the technology's development. Standards thus reduced the effort needed for mCHP appliances to pass the certification process. They reduced the need for extensive proofs of technical solutions meeting the essential requirements and provided a basis for a common understanding of these requirements:

> So for them [the manufacturers], it's easier that there is now a standard pointing clearly what the relevant [requirements] are.

In fulfilling this function in the certification process, standards supported mCHP's access to the European market and therefore played an essential role in enabling the technology's diffusion. While using standards remains voluntary and other solutions are acceptable, there was a widespread sentiment among the interviewees that adhering to standards related to the applicable European Directives (see Table 3.2) was almost a necessary condition for bringing mCHP technology to market and that other solutions should only be chosen in exceptional cases.

3.4   Standards' Additional Effects on mCHP's Development and Diffusion

Interviewees reported that the standards which were relevant for mCHP (see Table 3.1) had both positive and negative effects for their innovation activities. They emphasised the effects of standards on the certainty regarding regulatory requirements and certification (see Sects. 3.2 and 3.3). These aspects were a major focus of their activities related to managing standards (see Chapters 4 and 5).

In addition, the experts also reported other effects of standards on both the development and diffusion of mCHP. The positive effects named in this context include standards often being useful information sources; standards supporting access to complementary infrastructures (e.g. the electricity grid); standards allowing the industry to signal mCHP's benefits to other actors; and standards helping build economies of scales for the innovation. Negative effects on the innovation usually were perceived when standards were out-of-date or required standards were missing. These perceived effects were the basis for how actors in the industry managed standards in the case (see Chapters 4 and 5). We explain the effects that standards had in the case in detail below.

3.4.1   Support of Standards for mCHP's Development

Often standards served as useful information sources in the development of mCHP, not only about regulatory requirements and testing procedures (see Sects. 3.2 and 3.3), but also about other topics. Especially in technological areas where the companies had no previous experience, like safety mechanisms related to shortcuts and switching the device off in emergencies or measuring the amount of electricity produced, interviewees explained that they could make use of standards in their designs:

> For the new functionality, especially for the generation of electricity, of course, they were new aspects for us. (...) For the things which are only new to us but which are self-evident, you have to follow them. So then standards are a good help to show you what you have to do.

In addition, because *"experience that has accumulated over decades is behind standards, especially in the electro-technical and gas areas"* (translated from German) this information also supported more commonplace design decisions in the innovation process:

> When I do not need to ponder every time 'this material, this screw and this seal – may I or may I not?' This is definitely helpful. (translated from German)

A second way in which interviewees perceived standards to support the innovation was the role that they played in defining interfaces to link mCHP appliances with other elements, such as the electricity grid; electrical and gas installations in buildings; and communication between electricity producing devices (see Table 3.1). These standards have not only been providing technical information for the companies' NPD activities but also have been supporting the innovation's eventual diffusion by offering certainty for the industry and eventually customers that the appliances would work with other elements as intended and limiting customers' needed investment in changing elements like the gas installations in their houses. However, interviewees pointed out that, for important interfaces, this support was only available at later stages of mCHP's development because the needed standards did not exist at all (e.g. communication between electricity producing devices), or needed to be adapted (e.g. standards for internal wiring of buildings, see below), making these interfaces an issue to be considered in the management of standards (see Sect. 5.1).

In addition, standards also were described as supporting mCHP's diffusion by helping to signal mCHP's qualities and benefits to other actors, like consumers and governments. This particularly applies to the product standard (EN 50465) which also covers energy efficiency of the appliances and supports the requirements of the Energy Labelling Directive (see Table 3.2). EN 50465 includes a formula that allows calculating the energy efficiency of mCHP devices. This formula is intended to form the basis for determining an mCHP appliance's energy label, which the directive requires it to carry (although this formula was a major point of contention during the development of EN 50465—see Sect. 5.2.2).

Finally, standardisation supported the heating industry in reaching economies of scales for mCHP technology. By being able to rely on existing components from other products and standardising new key components, such as the Stirling engine, between manufacturers, the industry was able to reach higher production numbers much quicker than would otherwise have been feasible and thus bring the technology's costs down to make the price-performance ratio more competitive with other heating solutions and enable faster adoption in the market than might otherwise have been possible.

3.4.2   Hurdles to mCHP's Development from Standards and Related Issues

Standards sometimes also were seen as hindering the development of mCHP. Some standards contained requirements which were based on outdated assumptions and which were difficult to implement in the innovation or would have severely limited its value to users. For example, pre-existing standards for electrical installations within buildings were written under the assumption that there are only devices in a building that consume electricity but no electricity producing devices. These standards would have required substantial changes to a building's electrical installations to install mCHP appliances in existing buildings, thus adding to the technology's costs and making it less attractive to consumers in the crucial market for replacement of heating boilers in existing buildings. Another example of outdated assumptions underlying standards concerned test procedures fixed in a standard which may assume a certain device-architecture and specify the assessment of certain components of an appliance which may no longer be part of a new design and have been replaced by other components.

A second notable area where standards have been imposing requirements that the interviewed companies sometimes found difficult to fulfil in mCHP appliances is the access to the electricity grid:

> Standards can also be used to hinder technologies. The 'Network Code Requirements for Generators' is in many areas... I don't want to say designed to... but I say it makes it very difficult, in particular for small electricity generators. (translated from German)

Another interviewee described these requirements for generators as *"a real problem for small generators, because it now sweeps up any generator in Europe that is greater than 800 W in power output"*. One key example of a difficulty resulting from this network code is the requirement for dealing with changing network frequencies, which changed while mCHP was under development (see Sect. 5.2.1) due to technological developments in other realms. While it was traditionally required to switch an electricity producing appliance off in the rare cases when the grid's frequency deviates from the usual 50 Hz, the new rules required generators to be able to remain online and adjust their own frequencies in line with potentially deviating grid frequencies. This development posed substantial challenges for Stirling-based mCHP appliances:

> Now it wants you to operate things from 47 Hz to 52 Hz or something, so it's much, much broader than frequency swing, which is very difficult for a tuned Stirling engine, free-piston Stirling engine. In fact, we can't operate over that wider band.

Standards which imposed hurdles for mCHP in this manner required (sometimes extensive) action during the technology's development, either by adapting the technology or the standard, in order to avoid negative effects on mCHP's eventual chances of reaching large-scale diffusion in the market.

Although hurdles for mCHP's development sometimes arose from standards (the two examples above being the most notable ones mentioned by the interviewees), there was consensus between the interviewees that the most serious standard-related obstacles to the innovation actually resulted from the absence of needed standards (either completely or on a European level). The absence of the product standard (EN 50465) outlined in Sect. 3.1 was key for the development of mCHP and necessitated substantial efforts when the industry engaged in standardisation for the technology (see Sect. 5.2.2). In other key areas, such as the natural gas composition; exhaust emissions; access to the electricity grid; or financial compensation for energy that is fed into the electricity grid, standards only did (and to some extent still do) exist on the national but not the European level. The following quotes are three out of many in our interviews that address this issue:

> So, each country has its own requirements and when you go through them, then Germany has a certain standard which involves some protections that should be in. For instance (...) how to test if you are connected to the grid. (...) So, indeed, in the United Kingdom is forbidden what is required in Germany.

> And this feeding into the grid is something which I still do not completely understand. On the European level, a standard exists on this topic. This standard basically consists of a rather large number of national appendices. And it explicitly states that the respective connection requirements in the individual countries, or even regions and network operator environments

(...) must be taken into account. And this varies tremendously across Europe. (translated from German)

And then there are the specific parts, in particular for the flue gas evacuation. There, we have a European patchwork which cannot be outdone. (translated from German)

Such differences across countries meant that different versions of mCHP appliances needed to be developed and certified for each country where they were intended to be sold. This implied additional development effort and made it more difficult to achieve economies of scales for the components that needed to be adapted for the local versions. However, one interviewee at the European association of the heating industry pointed out that this might not be completely against the interests of the OEMs:

Honi soit qui mal y pense. Of course, the manufacturers do not want movement of goods to be as free as the consumer might think. There are also price differences between countries and they are thereby being blocked a little bit. (translated from German)

3.5   Overall Impact of Standards on mCHP's Development

In terms of their overall impact on the development of mCHP, interviewees saw standards mostly positive. Although there were some negative effects, as outlined above, there was consensus among the interviewees that these were by far outweighed by the positive aspects. This sentiment is represented by the following quote which characterises standards' function as proving a foundation for the innovation's development:

The aim of standardisation is very clear. At this moment, at this early stage of the technology, it is to lay a good foundation for this technology, so that this technology can be accepted by the market. (translated from German)

Based on the characterisations of support and hurdles arising from standards, they can be grouped according to (1) their link to regulation, and

Table 3.3 Standards' potential implications for mCHP

		Standard's link to regulation		
		Harmonised	*Linked to regulation but not harmonised*	*No link to regulation*
Innovation's ability to conform to standard	*Yes*	Type 1: Enabling market access and providing legal certainty	Type 3: Facilitating market access	Type 5: Facilitating product development
	No	Type 2: Effectively locking the product out of the market	Type 4: Complicating market access; affect product's position in the market	Type 6: Requiring own technological solutions

(2) whether the innovation can conform to the standard or not. While the first characteristic determines the strength of the impact on mCHP, the second characteristic determines whether this impact is positive or negative (see Table 3.3). Furthermore, several standards, which were needed to market mCHP appliances, did not yet exist when the technology's development started. While already existing supporting standards were relatively straightforward to manage, standards that hindered the innovation and/or were still missing required substantial attention during the technology's development. We portray these management activities in Chapters 4 and 5.

REFERENCE

European Commission. (2017, September 25). *Harmonised standards.* Retrieved from http://ec.europa.eu/growth/single-market/european-standards/harmonised-standards_en.

CHAPTER 4

Managing Standards for mCHP
on Company Level

Abstract micro Combined Heat and Power (mCHP) technology was developed by several established companies and start-ups in parallel. This chapter provides detailed insights into the different companies' innovation management approaches. Based on in-depth interviews, it compares how these firms managed standards and regulation while developing their mCHP products. It shows the types of awareness, expertise, and resources needed to provide a solid foundation for addressing standards and regulation that affect an innovation. Building on this, the chapter shows how these factors enable managers to introduce their innovations into highly regulated markets.

Keywords Innovation management · New product development Regulatory compliance · Standards · Regulation

The findings outlined in Chapter 3 show the importance of standards for developing the technology of mCHP and bringing the appliances to the market in Europe, thus making standards a key issue to manage as part of this development. Processes to manage these standards occurred on two levels: (1) Each of the involved companies had its own internal NPD process, as part of which standards were addressed. (2) In parallel to these company-internal activities, the industry collaborated on developing new and adapting existing standards to allow mCHP's development, where needed. Both levels interacted throughout the process, i.e. work

© The Author(s) 2019
P. M. Wiegmann, *Managing Innovation and Standards*,
https://doi.org/10.1007/978-3-030-01532-9_4

within the companies reflected the industry-level developments, and the activities to adapt standards were driven by the individual actors in line with their internal activities.

In this chapter, we focus on the company-level activities related to managing standards for mCHP (see Chapter 5 for a description of the collaboration between actors in the industry). There was a variety in approaches to managing standards and regulation and the degrees to which they were seen as important, as the following quote from an interviewee at a notified body illustrates:

> You see differences. Some manufacturers, they – I mean if we have this pre-assessment we push them to really read standards and then you see that some of them, they even haven't bought one.[1] And others, they already read it three times. So there is a difference in experience and seeing the need of using these standards.

We summarise these different approaches in Table 4.1[2] and outline them in more detail below. In Sect. 4.1, we focus on the companies' general approaches to standards and regulation. This includes aspects such as their awareness of the topic and the degrees to which it is handled strategically, as well as how standards and regulation are embedded into the companies' structures. Section 4.2 then shows how the interviewed companies incorporated standards and regulation into the mCHP development process, covering aspects like the timing of their management, how the companies identified relevant standards and how they incorporated input from the industry level into their development activities.

---

[1] Actors wishing to access the contents of standards developed by the ESOs and their national member bodies must buy the documents from the publishing arms of the standardisation organisations.

[2] We omit component suppliers from this table because all three interviewed component suppliers' activities related to regulation and standards were tightly linked to those of the appliance manufacturers, rather than standing on their own.

Table 4.1 Overview over appliance manufacturers' activities

Type of company	Company A *Established company*	Company B *Established company*	Company C *Established company*	Company D *Established company*	Company E *New entrant*	Company F *New entrant*
Awareness of standards' and regulation's importance	High	High	High	High	High, focussing on certification-related issues	Medium on certification-related issues, low on other issues
Technological expertise	High	High	High	High	High	High
Standardisation and regulation expertise	High	High	High	High	Low	Low
Available resources for influencing standards and regulation	Sufficient	Sufficient	Sufficient	Sufficient	Insufficient	Insufficient
Organisational structures for managing standards and regulation	No dedicated staff, the topic is coordinated by a senior engineer	Dedicated staff; additional company internal database of experts to support activities	Dedicated staff	No dedicated staff, the topic is coordinated by the head of the product certification department	Absent	Absent
Degree of strategic orientation for managing standards and regulation	High	High	High	Medium	Low	Low

(continued)

Table 4.1 (continued)

	Company A	Company B	Company C	Company D	Company E	Company F
Type of company	Established company	Established company	Established company	Established company	New entrant	New entrant
Core actors in identifying standards and regulation for mCHP	Company's own engineers	Company's own engineers	Company's own engineers	Company's own engineers	Notified bodies and consultants	Notified bodies and consultants
Conformity evaluation during development process	Important elements carried out in-house	Important elements carried out in-house	Important elements carried out in-house	Important elements carried out in-house	Heavy reliance on notified bodies	Heavy reliance on notified bodies
Timing of addressing standards and regulation in NPD process	In initial investment decision and at all subsequent stages of process	In initial investment decision and at all subsequent stages of process	In initial investment decision and at all subsequent stages of process	In initial investment decision and at all subsequent stages of process	Throughout process	Only in late stages of process
Participation in technology development collaboration	Yes	Yes	Yes	Yes	Yes	Yes
Participation in standardisation and regulation processes	Yes, in a leading role	Yes	Yes	Yes	No	No

4.1 COMPANIES' APPROACHES TO MANAGING STANDARDS AND REGULATION

As the quote in the introduction to this chapter shows, companies in the industry differ substantially on their fundamental approaches towards standards and regulation. Their awareness of the topic's importance varies (Sect. 4.1.1) and they are able to devote different amounts of the required expertise and resources to managing the subject (Sect. 4.1.2). As we outline in Sect. 4.1.3, these different foundations affect the grounding of managing standards and regulation, both in terms of strategic focus and integration into the organisation.

4.1.1 Awareness of Standards' and Regulation's Importance

A first factor driving companies' approaches to managing standards in the context of mCHP were the degrees to which they were aware of the topic's importance for developing the technology. This differed according to functions of standards and regulation, such as certification and providing market access, or acting as information sources.

4.1.1.1 Awareness of Standards for Certification and Related Issues

Standards and regulation can have a major impact on the certification, market access, and liability questions related to a technology like mCHP (see Chapter 3). One interviewee described this significance as follows:

> Both for the technology and the company – the success and the safety of a company – standardisation is an elementary topic. And companies and start-ups must be aware of this. (translated from German)

Most established companies acted in line with this view on standardisation and regulation. Based on their experience in the industry, they treated managing standards and regulation as a necessary condition for successfully developing new products and bringing them to the market. On the other hand, new entrants to the market sometimes did not understand the importance of standards and the European system, as the following quote from an interview with an engineer from a notified body, who had conducted conformity assessment of many companies' mCHP appliances, shows:

Basically, these boiler manufacturers, they already know standards, they know certification processes, so they were from that perspective better prepared. But on the other hand, the start-ups or the Japanese or the Americans are not familiar with the European situation. They were not that focused yet in standards, although some manufacturers were already (...) prepared but some of them were not prepared. Especially the start-ups – for them it's new to read and understand these standards, seeing the complete picture is difficult for them. And that's also the case for all parties outside Europe, they don't understand our system with directives and standards.

While none of the companies that we interviewed lacked awareness to a degree described in this quote, two of the smaller start-up companies explained that their awareness developed throughout the development of mCHP. When these two companies initiated their activities in the field, they did not yet know about the need for considering standards which caused some duplications of effort in the NPD process (see Sect. 4.2).

4.1.1.2 Awareness of Non-certification-related Functions of Standards

On functions which are unrelated to certification that standards can fulfil, such as providing useful information for the technology's development or defining interfaces, we observed more variation in the awareness among our interviewees. Interviewees at smaller companies mostly focussed their attention completely on standards which are related to certifying the product. They therefore did not seem to have a high degree of awareness of standards' other functions.

In established companies, interviewees were aware that standards can also fulfil non-certification-related functions. For example, interviewees brought up standards defining interfaces between a heating boiler and a building's pipework, standards providing information about characteristics of materials for certain applications, and standards reducing variety in components like control electronics. When these functions were mentioned, this was an aspect 'on the side', and interviewees saw them as a given when developing new products. They considered them such a basic element of their companies' internal innovation processes that they did not warrant much attention as part of managing standards and therefore these functions did not play a major role in the interviews.

Nevertheless, the non-certification-related functions of standards were significant for developing mCHP in the collaboration of parts of the industry that we describe in Chapter 5. Examples include reducing variety by standardising the Stirling engine component across different companies' products, facilitating collaboration in technology development (see Sect. 5.1.1 for both), and defining interfaces with the electricity grid (see Sect. 5.2.1). In addition, developing a standard to provide information about appliances' energy efficiency was a major focus of the industry's collaboration (see Sect. 5.2.2).

4.1.2   Expertise and Resources for Managing Standards and Regulation

In addition to a company's awareness, its available expertise and resources are key to the ability to manage standards and regulation effectively. As outlined below, we found in our interviews that this work requires specific expertise which can only be provided if a company has substantial resources at its disposal.

4.1.2.1 Required Expertise for Managing Standardisation and Regulation

Our interviews show two distinct topic areas in managing standards and regulation that require different types of expertise: (1) topics with technical, subject-related focus, and (2) topics on a higher, strategic level. The first area comprises all work that is directly connected to the technical contents of the standards, such as contributing to the development of technical requirements in standards and regulation, assessing their implications for product design, and implementing them in technical development. It therefore often requires in-depth subject knowledge. Tasks related to the second type include, for example, following ongoing developments in standardisation and regulation, assessing their significance for the company, and deciding whether and how the company should engage in standardisation and regulation initiatives. This also aims to coordinate the company's standardisation and regulation initiatives, e.g. in terms of assuring that input into a standard for one technology does not result in issues for another technology in the portfolio. One interviewee described his work in this context as follows:

> I am responsible for the strategic association work (...). Strategic association work distinguishes itself from operational association work because it is concerned more with which associations we should be part of: Where do we need to represent our interests and, if we have interests there, what

are our positions in the respective topics which are covered by the associations? (...) In addition to the strategic association work, the area of political lobbying belongs to association work. (translated from German)

In addition to the skill sets required for these distinct activities, interviewees agreed that effective of standardisation and regulation and representing the company in external working groups also necessitates staff with a high level of social skills, as the following quote shows:

It is equally important that one has the appropriate standing in these committees. Social skills in the widest sense. Because otherwise one leaves these committees with a lot of confusion and little results. (translated from German)

4.1.2.2 Required Resources for Managing Standardisation and Regulation

Providing the required expertise for managing standardisation and regulation is resource intensive. Especially in the early phases of a technology's development, many issues related to the topic must be resolved. There was consensus among interviewees that new technologies, such as mCHP, require substantial initial effort until the needed standards and regulation are established and all involved parties (manufacturers, notified bodies, regulators, market surveillance authorities etc.) are familiar with the technology. Once a technology has been established, the effort required for managing standards and regulation (e.g. following ongoing developments and contributing to keeping standards and regulation up-to-date) is much smaller.

Accordingly, interviewees reported using substantial resources for managing standards and regulation in mCHP's development. One interviewee stated that his company invested several man-years of work time into mCHP-related standardisation and regulation questions as part of developing the technology. Another interviewee estimated that the work of one out of approximately 30 full-time-equivalent positions involved in developing mCHP at his company was related to the topic. Overall, all interviewees whose companies participated in standardisation and regulation work estimated the effort to be somewhere between three and ten per cent of the overall time and effort for developing mCHP.

Standardisation—and regulation-related activities therefore comprised a relatively small but still significant share of all work needed to bring mCHP technology to the market. In larger established companies, these resources were usually available as needed, although one interviewee explained that it could sometimes be difficult to convince direct superiors of the required experts to make their staff available for standardisation work because the benefits may be long-term and/or difficult to measure.

Smaller start-up manufacturers explained that their limited resources sometimes hindered their ability to effectively manage standards and regulation, even if they were aware of the topic's importance. Especially participation in standard development and lobbying for changes to regulation was often unfeasible for them, as the following quotes show:

This [participation in standardisation], especially for a small enterprise, is very difficult. Such a new product development by itself already needs a great deal of resources and providing them in a company of our size is already, in my opinion, a considerable achievement. (translated from German)

Definitively, this [participation in standardisation] is an enormous advantage, clearly. But, as I already said, there always is a balancing act at our company regarding what personal and financial resources are available. If one wants to participate there, participate really constructively, then one also has to invest quite a bit. And for us, this is always a balancing act what can be used for that or whether our means can better be used in another place for the actual development work. (translated from German)

Unfortunately, they [the company's clients] didn't pay you to do that [participating in standardisation] and within [company name] we never had enough people. Again, this is where it's difficult to do a lot of product development and standards development from within a small company because we don't have the people, we don't have the money. Yeah, it would be nice to.

4.1.3   Strategic and Organisational Grounding of Managing Standards and Regulation

The degree of companies' awareness of standards and regulation and/or the available expertise and resources determined how the topic was grounded in the company's organisation. This in turn was linked to which degrees the companies could address the topic strategically. Some companies address these issues in an ad hoc manner whereas others have very clear structures and procedures for addressing standards and regulation.

The smaller start-ups we interviewed fall on the 'ad hoc end' of this spectrum. Their lack of dedicated resources meant that they were only able to address the most pressing standardisation and regulation issues at the point when they occurred and could rarely address the topic in a very strategic way. Other companies spent substantial resources to put clear structures in place that support managing issues related to the topic in a strategic and coherent manner. In between these two extremes, other companies implemented some elements to steer their standardisation efforts while using fewer resources to do so. We outline these observations in detail below, focusing (1) on the organisational structures for the management of standards and regulation, and (2) the intra-company networks to facilitate these activities.

4.1.3.1 *Organisational Structures for Managing Standards and Regulation*

In order to provide the skills needed to fulfil the tasks outlined in Sect. 4.1.2, the companies attached standardisation and regulation activities to different parts of their organisational structures. The first, subject-specific area of activities was directly linked to the product development activities for mCHP at all interviewed companies. It was often stressed during our interviews that it is essential for effective management of standardisation and regulation that a company's representatives have in-depth technological knowledge. The following are only a few of many quotes in the interviews which stress this importance:

> It is very important that in meetings where these topics [standardisation and regulation] are discussed, the technical expertise is present to talk about these topics, so that one does not just stop and say 'I am going to discuss this and come back next time' but that one is immediately in a position to make the required points. (…) Otherwise (…) one has to rework everything back at the company, [then] goes back [to the committee], but they are already further. This really hinders the process. Especially these technical expertise and social skills of those who work there and their internal network in the development departments is very important. One cannot simply send any – I don't want to say business economist – who is detached from the technology. (translated from German)

He [the company representative in standardisation] was extremely close to the project team [and] was very, very deeply involved in the development activities. This means it was not like we had a separate department which assumed the standardisation activities. Instead, the people who were very close to the project also did this. (translated from German)

It has always been important that one directly implements this experience which one has gained in [product] development in the standard. This is extremely important. This is also why the employees who have contributed to the standardisation committees – they all were employees from the new product development area. (translated from German)

And it can absolutely go so far that developers come along to, for example, the ministry of economic affairs to present a topic, explain a topic, precisely because these relationships are partly not trivial and are also not immediately accessible to civil servants, even if they have been at home in this subject area for a long period. Using development engineers for such communication tasks in our association work is something that we have been doing relatively often in the last years. (translated from German)

All interviewed companies assigned subject-related tasks in managing standards and regulation to the development engineers whose work already addressed these technological questions. In contrast, they differed regarding where in the organisational structure the responsibility for the more strategic questions was located. Specifically, we observed three different ways in which this was addressed: (1) Companies at the very ad hoc end of the spectrum of standardisation approaches did not address strategic questions at all, usually because of lacking awareness and/or resources. (2) In companies falling in the middle of this continuum, the topic was often covered as an additional activity by one or a few employees who were also otherwise involved in managing standardisation in regulation. For example, these tasks were handled in one company by a senior product developer and in another one by the head of the department responsible for product certification:

At [company name], we have a division which mainly occupies itself with certification, conformity declaration and so forth. And the head of this department dealt with the coordination [of standardisation activities] in close consultation with the development projects. (translated from German)

(3) Finally, two companies stand out because they have dedicated teams and can therefore be located at the very strategic and professional end of the continuum. The members of these teams to some extent also had a formal function to guide their companies in choosing where to engage and in defining common positions that should be followed by all staff representing these companies in standardisation and regulation. In the first example, the company established a team that is directly responsible to the head of product development which focuses on the strategic questions related to standardisation. In the second example, a team within the company's department of public relations is charged with these topics.

> I am responsible for the strategic association work (...). And we are embedded in public relations. (translated from German)

4.1.3.2 Intra-company Networks for Supporting Standardisation and Regulation Work

The organisational structures outlined above mean that the subject-specific questions are potentially addressed by many different experts. While some of the necessary alignment of their activities is ensured by the staff who address the strategic level of a company's standardisation activities, a consistent approach to standardisation also requires communication among the company's experts. In addition, some of the quotes above also show that there is a need for them to remain connected to other engineers who do not participate in standardisation themselves.

In several companies, we observed informal networks to ensure this communication. For example, we learned that one company's engineers who participate in standardisation keep each other informed about their activities through regular e-mail exchanges and other informal communication. Beyond such an informal approach, interviewees at a company that falls on the professional end of the standard-management-spectrum also explained that they support this intra-company network with a database which keeps track of all of the company's standardisation activities and the experts who are involved in this work:

> *Interviewee 1*: [We were talking] of the integration and transmission of information from mainly standardisation committees or maybe also associations into our company structure. For standardisation, we have a network where we can approach specific people through a matrix if

we have specific topics. (...) And in this network different people are named with different focus topics. And they are simply involved if you have such a topic. They then get the information.

Interviewee 2: This is the same for industry associations. (...)

Interviewer: This means a product development team can say 'we now have this problem here, we are now searching the database for the relevant person and approach him'?

Interviewee 1: This as well, exactly. [And] you can also share information between, I say, stakeholders who are located in different parts of the company. And they know through this (...) company internal network who has also dealt with this specific topic. (translated from German)

4.2   INCORPORATING STANDARDS AND REGULATION INTO mCHP DEVELOPMENT

Following our outline of the general approaches that the companies in the case took towards standards and regulation, we now describe how they incorporated the topic into their development activities related to mCHP. Because most of the interviewees focussed on standards that are relevant for safety and obtaining certification for their mCHP appliances, we also emphasise these areas in our description.

Our interviews reveal four core themes in this context: (1) identifying applicable regulation and standards (Sect. 4.2.1), (2) using them in specifying the company's product (Sect. 4.2.2), (3) evaluating the product's conformity to applicable standards and regulation (Sect. 4.2.3), and (4) the degrees of freedom for technology development afforded by standards and regulation (Sect. 4.2.4).

4.2.1   Identifying Applicable Regulation and Standards

In a first step of managing standards and regulation for mCHP, the companies needed to identify which regulatory texts and standards would be applicable to the technology's development. Doing so was important because companies entered new areas where they were unfamiliar with the requirements for the technology. In addition, regulation and standards are not static, meaning that the companies needed to stay aware of changing requirements. We observed two fundamentally different approaches to identifying applicable standards and regulation:

(1) an active approach used by the established companies, and (2) a more passive approach used by the smaller appliance and component manufacturers. Following an outline of these two approaches, we explain how companies in the industry anticipated changing and new requirements for mCHP.

4.2.1.1 Active Approach

Established companies usually started with an initial identification of areas of requirements that apply to the technology.

> At a very early stage when one defines the product specifications, it has to be clear which standards need to be fulfilled. (translated from German)

This involved the question which European directive(s) applied. Although the characteristics of the technology meant that a number of directives were already set for mCHP (see Table 3.2 for an overview), companies had some leeway in deciding which of them should be the *"leading directive"* (translated from German). All of the interviewed companies chose the Gas Appliance Directive for this purpose, due to their experience with previous products that had been certified based on this directive. This primary choice of directive(s) then guided much of the further search for standards. The following quotes from different interviews illustrate this approach:

> Before we address standards, one actually has to go a step back. Before one does this at all, one has to say in today's environment 'which directive do I even want to comply with?'. (...) And accordingly, I then have to look which standards are available. (translated from German)

> For us, it was clear relatively quickly that we want to work according to the Gas Appliance Directive. The Machinery Directive was also being discussed. But since we certify all our other appliances according to the Gas Appliance Directive, it was actually clear quite soon that we want to go in that direction. (translated from German)

> It always has been clear that the Gas Appliance Directive plays a role because the appliance will always have a gas connection, that the Low Voltage Directive will play a role because the appliance always will have an electricity connection, that the EMC Directive plays a role because the appliance has electronic components which can emit or receive electro

magnetic interference. These three directive are always a given, they are
also always a given for our current heat generators, you always have to go
by them. (translated from German)

The companies were already familiar with directives from their previous
products and they also knew most applicable standards in that context,
e.g. for gas safety. In other areas, e.g. related to the electricity producing
aspects of mCHP, a relative lack of knowledge and experience meant that
additional applicable regulation and standards had to be identified after
the initial search. In an iterative approach, the search for regulation and
standards was linked to the NPD process where moving on to new technological topics also led to the discovery of new standards and regulation
for mCHP. The following quote illustrates this:

> [At the time] we don't have any experience of or knowledge on electric-
> ity generation. So there you're treading a kind of 'terra incognita' and we
> have to find our way. We're discovering things – some from the outset and
> we see already at the beginning... 'How does that work with the grid?',
> 'How to connect with the grid?', 'And what are the requirements?'. And
> some [topics] we are discovering a bit later, for instance domestic wiring.
> So, it's a mix in fact of thinking ahead and discovering while you're going
> your way.

4.2.1.2 Passive Approach

Smaller companies relied to a large degree on other parties to identify
the applicable requirements for their products. For example, the interviewed start-up appliance manufacturers used the support of notified
bodies and/or consultants:

> *Interviewee*: At this point [...] it was about standards and which standards
> we have to comply with. And then we hired two consultants, one in
> [the country where the company's R&D department was based] and
> one consulting company in the Netherlands. This consultancy company
> is [name of a notified body].
> *Interviewer*: And they in essence created a kind of list for you of the stand-
> ards that were relevant for the topic?
> *Interviewee*: Exactly. And at this point they have accompanied us very well.
> (translated from German)

> *Interviewee*: We had to find out for ourselves first which standard – if we wanted to have the mCHP appliance tested as a whole with the aim to obtain a CE-mark – which one would apply there at all.
>
> *Interviewer*: And how did you proceed to determine what applies in this case?
>
> *Interviewee*: On the one hand we got in touch with the test laboratories which are active in this area and discussed with them according to which standards they would conduct the tests or which standards apply according to their opinion. And then, in parallel, we also conducted our own search based on these insights. (translated from German)

This role of the test laboratories was confirmed by our interviewee at a notified body:

> The process starts very often with the, we call it pre-assessment meeting, where we (…) discuss (…) the complete overview of relevant standards.

Component suppliers also used help from external parties. Because component suppliers were mostly not directly involved in the certification process, they largely relied on the appliance manufacturers to inform them about the requirements arising from regulation and standards. The following quote illustrates this approach:

> When this specification sheet is created (…) these are on one hand market requirements (…) but of course also legal requirements. Especially for gas and electricity there are clear safety requirements that must be fulfilled. There is no way around this. The thing is that we get this from our cooperation partner – because he is responsible for bringing [the appliance] in circulation – in a relatively nicely condensed way from one source. That makes it easier. (translated from German)

This reliance on appliance manufacturers to provide lists of applicable standards is partly explained by their ultimate responsibility for the entire product's safety but also by their better knowledge of the application area. For example, one fuel cell manufacturer supplied fuel cells to both mCHP and automotive applications. Our interviewee at that company noted that the standards and regulation in these areas differ to a large extent, making it difficult for suppliers to stay up-to-date and understand the specific requirements without their customers' support.

4.2.1.3 Anticipating Future and Changing Requirements

In addition to identifying current standards and regulation for mCHP, companies in the industry also needed to anticipate future requirements for the technology:

> If suddenly any new requirements, which impact on our development, come out of the standard, then it is extremely important to know this at an early stage. (translated from German)

Because mCHP's development took several years and the products needed to be certified according to the requirements in place at the time when they were released to the market, it was essential to already anticipate these requirements during the design process. Participating in standardisation and other working groups is key for learning about—and influencing—these developments (see Chapter 5). In addition to information about upcoming standards and regulation, this participation also provided the companies with further knowledge. In many cases, participation in standardisation committees brought them in contact with stakeholders outside the heating industry. This provided insights into these stakeholders' needs, their views on mCHP, and implications for the products' design in order to make the technology acceptable for these external stakeholders and even provide additional value for them (e.g. in the context of electricity grid stability, see Sect. 5.2.1).

While much of this information about upcoming requirements and other stakeholders' views was obtained by participating in standardisation, the participation's resource intensiveness sometimes made this unfeasible. Established companies sometimes relied on external consultants who participated in standardisation committees on their behalf whereas the smaller companies again largely relied on notified bodies to obtain information before new standards and regulation were made publicly available:

> At this point we have, for example, a consultant who informs us, for example, about technical standards. Through this pipeline, through this consultant we get tips about which new standards are changing for us now and in the future. And as a second channel, [name of notified body] informs us about changes. (translated from German)

Especially for the smaller companies with insufficient resources, this was the only way of accessing advance information about upcoming standards, putting them at a disadvantage compared to established players who could directly participate in the process or hire consultants to do so on their behalf:

> Of course, we always got access to this [information about developments in standardisation] a bit later. This is clear. I would say that there have been tips from time to time in which direction this goes or similar things. But this is, as I already said, a process which you have to accompany continuously if you want to be really close to it. And this does not always work when you also have to deal with every-day problems. (translated from German)

4.2.2   Specifying the Product

Following the identification of requirements for mCHP, their implications for the product needed to be specified. This specification of the requirements had far reaching consequences for mCHP's further development, the product's viability, and thus eventually also the technology's success. A first step in specifying the requirements was 'translating' them into concrete technical terms and including them into the product's specification sheet, which took substantial effort in itself:

> We had requirements from the standards but the process [within the appliance], the appliance, the concept must first undergo a risk analysis from which requirement specifications are derived: 'What do the controls look like? Which sensors are required? What is the performance? Which failure models?' (translated from German)

As part of this activity, the established companies[3] also faced the question whether to apply the existing standards and regulation to the technology or whether to attempt influencing the requirements (see Chapter 5 for a description of how they did do so):

---

[3] The smaller start-up players did not face this choice due to their limited resources, and had to design their products based on the given standards and regulation.

You have the product and you have the regulations and finally they have to comply, either by changing the product, adapting the product to the regulations or by adapting the regulations and standards to the product.

4.2.2.1 External Support for Specifying Requirements

Because of the importance and complexity of specifying the requirements, most interviewed companies again called on external support, like they did in identifying the requirements. This support came from (1) notified bodies, (2) external consultants, and (3) using pre-specified components.

Again, the smaller start-ups relied on notified bodies' help to understand the contents of relevant standards and regulation. Their consulting activities accompanied these players' development of mCHP products and included an important element of explaining the requirements:

> We started with this pre-assessment, then the consultancy phase, to assist them in understanding the requirements and the standards.

> Our consultancy is really focussing on the standards, on the content of the standards.

Although the notified bodies performed such consulting activities, these activities were limited in scope and could not cover the full specification process in order to avoid conflicts of interest when eventually certifying an mCHP appliance. The notified bodies could not go as far as proposing design solutions or supporting the companies' risk assessment, which were assessed at a later stage in the certification process. This made some of the notified bodies' consulting work as 'grey area', as our interviewee at a notified body acknowledged, and they needed to be careful not to exceed their role:

> Of course, there is a grey area. (...) We cannot do a risk assessment of an appliance because afterwards we have to assess this risk assessment. That's not allowed, so the consultancy we do is advising them on the requirements in the standards. (...) So, we give them some guidance but we cannot say 'you have to change this'. That's not our role.

Because of these limits to the support that the notified bodies could provide, several companies, including all major actors who we interviewed, also relied on an independent consultant in the field. Several interviewees named him as the leading expert for standards and regulation for mCHP. This consultant described his focus as "*consulting companies during the development of a safety-related concept*" (translated from German). He was involved in various ways in the product development of the different companies to support them in implementing the standards and regulation. Sometimes he was involved only at selected points in the companies' NPD processes to address specific issues, e.g. when notified bodies pointed out problems during the certification process that the companies could not address without help. In other cases, his input into technology development was much more substantial:

> My development work in many of these projects is writing the safety-related specifications of the requirements. There you write in detail: 'Which standards, which features and how are they implemented?' In some cases, I also write the safety-related concept for the software. (...) My consulting goes up to successful certification. (translated from German)

In addition to hiring external experts for support in the specification process, companies could also rely on pre-specified components from suppliers for certain safety-critical parts of the appliance. Especially smaller companies made use of this option. This allowed them to meet key requirements from standards and regulation without spending scarce resources on own developments and specifications:

> There are certain safety devices. This is, for example, the automatic firing device which we do NOT develop ourselves. This is a purchased part from companies like [company names] which have been established in that area for years. These developments cost a lot of money because they include building failsafe controls and software. They are inspected by a notified body and we then rely on ready-made products. We cannot afford to develop such things ourselves. (translated from German)

4.2.3   Evaluating Conformity to Regulation and Standards

In order to make their final products conform to the regulation and standards, companies also needed to evaluate this conformity at different stages in the development process. Below, we outline what we learned about (1) the initial evaluation at the outset of their development projects, and (2) the review procedures throughout the development process.

4.2.3.1 Initial Evaluation of Regulation and Standards for mCHP

Especially the established companies, with their high awareness of regulation and standardisation and their professional approach to managing the topic, already addressed standards and regulation as an issue in their initial appraisal of mCHP technology's potential. When making the business case for mCHP and deciding whether to invest in its development, an analysis of the degree to which standards and regulation would support or hinder the technology was essential:

> A certification capability analysis, doing this is a standard procedure. Is this product even capable of being certified at all? Are there any hurdles from a standard or regulatory point of view? This is something one does very early. (translated from German)

Such evaluations often did not only consider regulation and standards that were directly relevant for certification but also could be wider in scope. The following example shows how important such analyses can be: One interviewed company first assessed the technology's potential in 2000 when it was concluded that the regulation for feeding electricity into the electricity grid was unfavourable, only allowing an insufficient return on investment for buyers of mCHP appliances. Because of this insight, the company decided not to invest in developing mCHP technology at that point in time. The company then re-evaluated mCHP technology in 2004. At that time, the requirements had changed and it was deemed feasible to manage remaining issues during the NPD process so that regulation and standards would no longer hinder mCHP when the technology would be ready for market entry. Following this assessment, the company initiated its development activities.

4.2.3.2 Evaluating Conformity Throughout the NPD Process

Following the decision to initiate the NPD process for mCHP, most interviewees stressed the need to assess regularly whether the developed solutions were in line with requirements from regulation and standards. At most interviewed companies, this was incorporated into the project management tools used to manage mCHP's development, e.g. by including the topic in the progress evaluation at regular milestones or in the companies' stage-gate processes. Doing so was seen as a way to prevent duplication of effort that would have been caused by not addressing the issue throughout the process and then having to adapt the product in the late stages of development to make it acceptable for certification and market introduction.

In several instances, the ongoing evaluations of conformity throughout the NPD process were also advised by the notified bodies and the independent consultant mentioned in Sect. 4.2.2. Especially the smaller players relied on the advice of notified bodies to identify areas that they needed to address before their products were ready for the certification process, as the following quotes from interviews with a start-up and a notified body show:

> We definitely tried to develop the first prototype in 2004 in a standard-compliant way. We also collaborated with a test laboratory which supported us in a consulting manner but we did not really try to get the CE-mark yet for this prototype because it was clear that we still would need fundamental revisions. (translated from German)

> And after that [the initial pre-assessment meeting] we dig into the technology itself and we check for what the risks are and where some parts of the system do not meet the standards, so the safety – this is purely focussing on safety. And then what follows is very often a kind of consultancy phase where they are further developing the system.

> So they say 'we have this safety concept' (...) and then we say 'OK, it does fit for 90% and this 10% does not fit'.

4.2.4   Degrees of Freedom for mCHP's Technological Development

A final theme related to managing standards and regulation in mCHP's development that recurred in our interviews was the degrees of freedom that the requirements left for developing innovative solutions. As we outlined in Sect. 3.2.2, not following standards carries substantial additional effort for the NPD process. Although *"undertaking this effort"* can *"sometimes [be] worthwhile if one has corresponding cost savings"* (translated from German), it became clear during our interviews that companies rarely did so in developing mCHP. Usually, standards were perceived as leaving sufficient freedom to develop the technology, and notified bodies were flexible in interpreting them, as the following quotes show:

> Standards usually leave the latitude to get equivalent solutions accepted – this is often the case. (translated from German)

> [Name of notified body] in this context paid attention to the content of the standards and not the wording of the standards. So the content – safety – was more important than narrowly [following the standard word-for-word]. Our engineers enjoyed the product-oriented interpretation of standards. (translated from German)

Despite this generally positive view on standards and regulation across all interviewees, we did observe some disagreement on two aspects related to how they should best be handled in the NPD process to provide optimal freedom for the innovation. This disagreement concerned (1) dealing with the missing standards, and (2) the timing of involving standards in the NPD process.

4.2.4.1 Handling Missing Standards in the NPD Process

As outlined in Chapter 3, some important standards for mCHP were missing when the industry started the technology's development and key requirements were therefore unknown at the outset of mCHP's development. Some of the interviewed companies saw the resulting uncertainty as a bigger problem for the whole NPD process. They therefore focused their efforts (see Chapter 5) on creating certainty as quickly as possible by engaging in standard development. However, other companies valued this situation as an additional degree of freedom for the engineers in developing the technology. They took this opportunity to experiment with new approaches to product safety, which they later contributed to the standardisation process:

> *Interviewee 1*: To the contrary, we could shape the standards very well based on our experience and the freedoms which we had [when the standard was still missing]. Especially not being regulated, overregulated and restrained too much in the beginning gave us much space to develop our safety concepts and develop ideas that we might not have had if there had been a relatively fixed standardisation frame. And this was very positive. As this point, we started using HAZOP analysis (...) a very interesting tool which we got to know in the USA and then

brought to Germany (...). And this is now also anchored in the standard. (...) And this has helped us a lot to be certain that we are on a good way with this new technology.

Interviewee 2: In collaborating with the Americans (...) – they had a different safety philosophy. (...) And with the standard as we have it now, there is on one hand clearly the European strategy of prevention but through the risk analysis we now have a bit more free space. (translated from German)

4.2.4.2 Timing of Handling Standards and Regulation in the NPD Process

A second aspect related to freedom for product development where the views diverged was the question at what stage in the development to start addressing questions related to standards and regulation. In particular one interviewee stressed that doing so too early would restrict the ability develop novel solutions, and that standards only became helpful at a later stage in the process when the prototype-mCHP-appliances were transformed to production models:

He [the manager of the development process] attached great importance at this point to avoid restricting the innovation through standards. They [the development team] perceived this as hindering in the early stages. (...) At this point in time standards would have hindered the engineers. (...) And then, at this point [later in the process], there is a bridge when the engineers see the need to be standard-compliant and this is helpful to bring the product to the market. (...) At this point, the company is getting used to standardisation and thinking in standards. When you standardise, when you produce in large numbers then you have certification, then you must [adapt] processes (...) and at this point, the freedom of the engineers is limited anyway. (translated from German)

[The development team] always (...) wanted a development strategy which put the innovation, the innovative element first. This is the fundamental thought which brings the product to life. And in this place, they always [aimed] to first find the technical solution and (...) later adapt it to the standards. Because you don't get a working system just like that and it can happen that a new development dies on the workbench in the lab if you already restrict it with standards at this stage. (translated from German)

In contrast to this strong view, all other interviewees advocated addressing standardisation and regulation early in the development process, as demonstrated by the very early first assessment of requirements outlined in Sect. 4.2.3 and shown by the following exemplary quotes:

> *Interviewee*: It's really important that with your first step this pre-assessment [involving the notified body] takes place in a very early stage of the development.
>
> *Interviewer*: So, is there already a prototype or even before that?
>
> *Interviewee*: Even before that is better. But in practice, I think, half of the cases, they already have a prototype. And some are very late. But I think about half of the parties, they didn't have a prototype yet, only paperwork.
>
> *Interviewer*: What would you suggest in general to a company in a similar situation which also develops a product where standards and regulation are relevant?
>
> *Interviewee*: Deal with this topic early on. (…) Not just developing a product or anything and then we'll see what we have to adhere to. Instead, incorporate this from the outset and say 'this is what I want to develop, what do I have to take into account?'. Not just having the technical specifications in mind but also looking immediately at what [requirements] are coming from the market and what we have to consider to bring it into the market at a later stage. (translated from German)

The interviewees, who favoured this approach of addressing standards early, reasoned that this avoided duplicate effort in developing the technology. According to this reasoning, the limitations in freedom for innovation imposed by standards only restrict the development of solutions that are not suitable for certification and therefore would need to be replaced by other approaches at later stages anyway (or require changing the standards). This is also reflected in the experience of one interviewee whose start-up encountered substantial rework in its early technology development projects because of not considering standards and regulation early enough and changed its development approach based on this experience.

CHAPTER 5

Industry-Level Collaboration in mCHP Standardisation and Regulation

Abstract This chapter provides in-depth insights into the extensive collaboration across multiple actors in the European heating industry during micro Combined Heat and Power's (mCHP) development. Actors in the industry cooperated both in developing mCHP technology and related standardisation/regulation processes. The chapter outlines the role of non-company actors (e.g. industry associations) and the industry's intellectual property rights approach (IPRs) in facilitating this cooperation. This chapter gives a detailed account of the particularly dynamic and contentious processes of standardising and regulating access to the electricity grid and requirements for energy efficiency labels. These examples show how innovators can jointly create conditions that support their innovation, even if major stakeholders (including government) oppose the technology. The examples also show how innovators can handle important policy and societal issues.

Keywords Cross-company collaboration · European Commission Energy efficiency policy · Electricity grid access · Intellectual property rights · Co-opetition

In addition to the internal activities described in Chapter 4, the actors in the industry also reached outside their companies as part of managing standards and regulation for mCHP. This resulted in extensive collaboration between actors in the industry. In Sect. 5.1 we

© The Author(s) 2019   77
P. M. Wiegmann, *Managing Innovation and Standards*,
https://doi.org/10.1007/978-3-030-01532-9_5

provide an overview of these activities, outlining aspects like the venues where this collaboration took place, the involved actors, the topics of cooperation, and how intellectual property rights (IPRs) were considered in this context. In Sect. 5.2 we then describe how standards and regulation for mCHP evolved as a result of this collaboration and the input of other stakeholders, based on two examples that were central to the case.

5.1   Collaboration Across Actors in the Industry

Having identified standards as an important issue for the development of mCHP, the actors in the industry also recognised that successfully bringing mCHP to market would be very difficult if companies tried to do so without collaboration in the industry. For example, the conflicts, which we describe in Sect. 5.2, would have been extremely difficult to resolve by any company from the industry on its own. This awareness resulted in extensive collaboration within the industry, both to develop the technology and its market, and to pursue standardisation and regulation-related activities together. This collaboration took place in a number of formal and informal settings with different aims and varying involved parties, many of which engaged in multiple collaborations with others. Table 5.1 provides an overview of the most important collaborations that were mentioned in our interviews.

We outline these collaborative efforts in more detail below. We first consider the initiatives which were specifically initiated for mCHP and included aspects related to technology development, but also standardisation and market development for the technology (Sect. 5.1.1, the four rows at the top in Table 5.1). We then outline the efforts in already established forums (concentrating on industry associations) which focussed much more on standardisation and regulation instead of technology development (Sect. 5.1.2, the two rows at the bottom in Table 5.1). These efforts led to some interesting 'group dynamics' between actors in the industry which we outline in Sect. 5.1.3. Finally, such collaboration also raises the question how the involved actors handled intellectual property. We take a closer look at the approach to this topic in Sect. 5.1.4.

Table 5.1 Overview of collaborations related to mCHP technology

Organisational setup of collaboration	Forum for collaboration	Aims of collaboration
Consortium, specifically initiated for mCHP	Initiative Brennstoffzelle (IBZ)	Promote and jointly develop fuel-cell-based mCHP, organise large-scale field trials of the technology
Ad hoc agreements between participating companies	Collaboration between a Japanese fuel cell manufacturer and a German appliance manufacturer	Jointly develop fuel-cell-based mCHP appliances for the European market
	Collaboration between several appliance manufacturers and a manufacturer of Stirling engines	Jointly develop Stirling-based mCHP technology and prepare the market for the technology. Later, the appliance manufacturers invested in the supplier involved in this cooperation
	Various one-on-one collaborations between appliance manufacturers and suppliers	Jointly develop components and other aspects of the technology
Established industry associations	European and national industry associations (e.g. EHI, COGEN Europe, BDH)	Provide a forum to coordinate the industry's input in standardisation committees and a channel for the involved companies to influence regulation for mCHP
Formal standardisation activities	Standardisation committees in European and national SSOs	Develop standards to support mCHP

5.1.1 Collaborating in Technology Development

Collaborations to develop mCHP technology began already in the early stages of development before the engagement in standardisation started and took place in settings that were specifically established for mCHP. Throughout our interviews, many instances of collaborating with suppliers and others to develop components were mentioned. Three of these technology development collaborations stand out because of their links to market development, standardisation, and regulation: (1) a collaboration between a Japanese fuel cell manufacturer and a major established

German OEM; (2) a German industry forum for domestic fuel cell applications and two associated field trial projects for mCHP appliances; and (3) a collaboration between several parties to develop Stirling-based mCHP technology.

In the first example, a Japanese manufacturer of fuel-cell-based mCHP appliances brought its extensive knowledge of the technology into the partnership. While this manufacturer produces entire mCHP appliances in Japan (where the technology has already reached widespread diffusion), it partnered with a German appliance manufacturer because of its limited knowledge of both European market requirements and European regulation and standards for mCHP. In this partnership, the Japanese company supplies the fuel cell components which are integrated into the appliance by the German appliance manufacturer who also has been responsible for questions related to standards and regulation.

In the second case, the German industry forum ('Initiative Brennstoffzelle', IBZ) brought together a large number of mCHP appliance manufacturers and other stakeholders, including academic research institutes, utility operators, industry associations, and a German government body in charge of promoting fuel cell technology ('Nationale Organisation Wasserstoff- und Brennstoffzellentechnologie', NOW). Its aims included information exchanges between actors, raising awareness for the technology but also developing technical specifications and political lobbying for the technology (see also Initiative Brennstoffzelle, 2017). The IBZ also had links with two large field trial projects ('Callux' and 'ene.field') which aimed to gain experience with the technology and testing prototypes in the field, but also linked to standardisation and regulation. The field trials relied on standards (e.g. for communication between the involved appliances), and produced findings that fed into further standardisation efforts later on.

The third major collaboration in the case aimed to develop Stirling-based mCHP technology. It involved the major appliance manufacturers which pursued the technology (although some of them have stopped their engagement before bringing Stirling-based mCHP appliances to the market, see Sect. 2.2.2). This collaboration took place in the early stages of development, as the following quote shows:

> In the beginning, meaning before our actual product introduction phase, we developed this Stirling engine together with competitors, mainly with

two competitors from the European industry. And then at some point we separated, so these common meetings eventually did not take place anymore. (translated from German)

In addition to the appliance manufacturers, a manufacturer of Stirling engines has been playing a key role in the collaborative development of Stirling-based mCHP appliances, being *"very deeply involved in that process, from the very first contact with [name of one OEM] right through to them producing and certifying their first model"*. In this context, the manufacturer not only developed the Stirling engine as an individual component but also was involved in integrating it into the appliances. This collaboration between the appliance manufacturers and the manufacturer of Stirling engines culminated in the appliance manufacturers jointly buying the Stirling manufacturer together with an external investor when the original owner (a large utility firm) decided to leave the mCHP appliance business.

One important motivation for this close cooperation between competitors was increasing the speed at which economies of scale could be reached for mCHP technology. The collaboration allowed them to standardise new components that were not shared with other products, such as the Stirling engine component or control electronics, across manufacturers. In addition, considerations about creating the market and being able to manage standards and regulation were further reasons for this collaboration. An interviewee at the company that initiated this collaboration explained why they decided to share their innovation with others, rather than protect it through patents and licenses:

> We were also active at that time to enlarge the circle of companies coming with micro CHP. So, we invited competitors because we thought it would be good that, when you have to create a new market for a new kind of product – If it is only the product of [company name] then it would be very much like the regulations had to be tailor made for [company name], for one company. And that was not the issue if it was for a sector. So, we collaborated with these different companies – also in lobbying on the regulations.

This sentiment of needing to collaborate in order to jointly develop the technology and the environment in which it is placed was also echoed by other interviewees, as the following quote shows:

If I had tried to distinguish myself from a competitor in this way and I wanted [...] to prevent him from implementing his technology – that would be absolutely counterproductive. The market first has to develop. The market for mCHP is not developed yet. It is a small plant and it needs to be watered well for it to start growing. (translated from German)

Based on these initial technology development efforts with their links to standardisation, the industry also engaged in established standardisation bodies and industry associations to further coordinate their activities in standardisation and regulation processes, as detailed in Sect. 5.1.2.

5.1.2   Collaborating in Standardisation and Regulation

In addition to the technology-focused collaborations outlined in Sect. 5.1.1, which also affected standardisation and regulation to varying degrees, there were a number of collaborative efforts directly concerning standardisation and regulation. They took place in different forums, such as the IBZ; the national and European industry associations[1]; and standardisation committees which were only *"one part of the network surrounding this technology"* (translated from German).

While there also was collaboration in the standardisation committees, it is particularly interesting to consider how collaborating in already established industry associations supported the industry's standardisation activities and provided the actors with access to regulatory processes. Especially the established appliance manufacturers engaged in the mCHP working groups at the industry associations but also some smaller players were members. By using the opportunities that these working groups provided, the industry was better able to cooperate in pursuing standardisation and regulation for mCHP beyond what would have been possible by only engaging in committees. Below, we outline how they used their membership in these associations both in the context of (1) standardisation and (2) regulation processes.

---

[1] These associations included the 'Association of the European Heating Industry' (EHI) and the 'European Association for the Promotion of Cogeneration' (COGEN Europe) on the European level and the 'Bundesverband der Deutschen Heizungsindustrie' (BDH) on the German national level.

5.1.2.1 Industry Associations in the Standardisation Context

Several interviewees reported that the actors in the industry used the associations to develop a common position which they could then pursue in standardisation committees, making them a venue to jointly prepare standardisation activities. For this reason, the companies were often represented by the same people in standardisation committees and the industry associations' working groups:

> It is often the case that there is an overlap of around 70% in people, who are on one hand active in standardisation topics and on the other hand in topics related to the associations. Yes, I would say that between 50% and 70% of these people are identical. (translated from German)

In order to facilitate this process, a representative of the European heating industry's associations participated in many relevant standardisation committees as an observer without voting rights. This allowed him to identify potential areas of conflict and facilitate compromises between the association's members in these areas. He also saw it as part of his role to ensure that the interests of smaller companies in the industry, who were not directly represented in standardisation committees, were also taken into account in these agreements. In instances when these interests were at threat in the committees, he intervened in the discussions. The following excerpt from an interview sums up this role:

> *Interviewee:* In the expert group, where the standard is being drawn up, only experts are present. This means that everyone has the same weight and everyone may speak or not speak – whatever they want. And I have been nominated as an expert. Of course, I hold off when members [of the association] voice specific demands. But if one member, for example, wants to push through certain things vis-à-vis other members of our association, then I have to intervene and say 'no, no, just a moment, there we have to find a compromise' because everyone sitting at the table, all members, must be able to survive. It cannot be allowed that someone raises a demand, let's say for example all appliances must be green, and the others want to have green, blue, pink. […] Then I have to intervene and say: 'No, no, that's not how it goes. Let's see whether we can leave the question of colour fully open.'
> *Interviewer:* Good, this means that, if that were the case, this member would have to go into the standardisation committee itself and say there 'we want green' and not through the industry association.

Interviewee: Yes, or he is sitting in the committee and demands this. Then I have to say 'no, no, that's not how it goes'. There are two ways.

Interviewer: This means you also counter this in the committee and say 'the consensus in our association is that we do not want to commit to anything here.'

Interviewee: Exactly. And if absolutely no compromise is found we go back to our internal working group and resolve the situation there. And usually this works out. (translated from German)

This role of the industry associations was mostly appreciated by the interviewed companies although a few clashes on minor topics with the association's representative were mentioned by one interviewee. This may also have been related to the representative working for both the German national and the European industry associations, making it sometimes unclear for actors from other countries on whose behalf he was speaking. In addition to these activities related to facilitating compromise and finding common positions for standardisation, the associations played one more role in standardisation for mCHP. Their staff also attended standardisation committees on topics which did not warrant the manufacturers' participation but were nevertheless relevant for mCHP and reported back on progress in these committees.

In some (mainly electrotechnical) areas of standardisation that were important for mCHP, this collaboration went even further than only agreeing on common positions for standardisation. In technological fields where actors in the industry sometimes lacked the necessary expertise and direct participation in standardisation would have been too resource intensive, they hired an external consultant through an industry association to act on their behalf in standardisation committees[2]:

There is an international standardisation committee where a strong electrotechnical aspect was included. There, we are not directly involved, but only through a consultant who we have mandated, together with our competitors, to represent our interests there. Doing this, with meetings in Tokyo and I don't know where else, is of course very resource intensive. This is why Mr [name of the consultant] is there. And Mr [name of the consultant] is paid for not by us as [company name] but by us as industry to represent our interests in international standardisation. (translated from German)

[2] The same external consultant also worked for many of the companies individually (see Sect. 4.2).

An additional reason for choosing the external consultant, rather than a member of the association's working group, to represent the entire industry was his neutrality resulting from having no links to a particular company:

> I was approached whether I could represent these bundled interests. It was also clearly said that it is better if a neutral non-producer of appliances does this instead of an appliance manufacturer. (translated from German)

5.1.2.2 Industry Associations in the Regulation Context

While engaging in the industry associations was (partly) complementary to directly participating in standardisation committees, it played a much more central role for the manufacturers in order to gain access to regulatory processes. This access was needed in particular when developing a calculation method for energy efficiency (see Sect. 5.2.2).

With the exception of one appliance manufacturer which is part of a larger conglomerate that operates its own substantial lobbying presence at the EU level, none of the actors in the industry would have had much clout in policy making on their own.[3] While the European Commission and other policy makers could be accessed by individual companies at industry roundtables and similar consultations about new regulation, the existing contacts of the industry associations helped to get more direct access:

> I think first they [the industry associations] know the way, they are close to the process, so they know what happens, they have the contacts already and so this is how this usually works indeed. [...] I must say, I have also been to – sometimes the European Commission themselves are organising a kind of round table meeting where you can register yourself. I have also been to that meeting but then there were 25 people in too small a room, and no individual talks.

In such instances, when members of the industry got access to policy making through the channels of the industry associations, they did so after a common position had been determined between the members of

---

[3] This manufacturer's ability to use its parent company's lobbying resources contributed to some interesting dynamics in the development of energy efficiency standards for mCHP, as outlined in Sect. 5.2.2.

the associations' working groups. They were then speaking on behalf of the entire group, also reflecting the reasoning for collaboration quoted in Sect. 5.1.1:

> The first time I was there [at the European Commission], that was through EHI – also with other people – and representing EHI. I've also been there later when EHI and COGEN Europe joined forces. I was there on behalf of and also together with people of EHI and COGEN Europe. So the general secretary of EHI was there, a colleague of [name] was there, […] the general secretary or director of COGEN Europe was there together with someone who was responsible for micro CHP and I was there.

In particular the interviewee who initiated much of the collaboration in the industry, and also was described as the leading force behind many of the common activities by others, was chosen to represent the industry together with staff of the associations (and—in some cases—additional external experts who were jointly hired by the industry) in this manner.

5.1.3   'Group Dynamics' in the Industry Resulting from the Collaboration

All interviewed parties who were involved in the collaborative efforts outlined above described them as very trusting. This trust was built throughout all of these efforts (i.e. technology cooperation, standardisation activities and collaboration in consortia and industry associations). The following quote from our interview with an academic engineering researcher, who participated in the process without commercial stakes and therefore played a more neutral role, sums up this sentiment:

> The nice thing about standardisation is that one tries there to work together and not against each other. This means that the idea of competition is secondary in a standardisation committee once the door closes. Evidently, everyone represents the interests of their company. This is clear. Nevertheless, one knows 'okay, one somehow has to enter compromises', otherwise nothing comes out and one eventually wants to have something on the table. This is similar to conducting a common research project where it is clear that one enters the whole thing as partners and tries to do something together. And this is the same in standardisation, at least in the micro CHP area, where – according to my experience – there are fewer

conflicts and diverging positions. Instead, the industry is saying – especially at such a new technology – 'okay, we pull together and we want to advance our niche products and our not yet established technology'. (translated from German)

This was sometimes also described as resulting in strong 'group dynamics' where all involved actors know each other very well and it may be difficult for outsiders to join these efforts. Some interviewees also saw these collaborations not only as a way to facilitate mCHP's development but also to fend off demands for requirements in the standards which would have been problematic for the technology. For example, one interviewee mentioned NGOs who participated in standardisation committees and who tried to raise the minimum levels for safety and exhaust emissions in the standards to such a high level that the industry would not have been able to produce mCHP appliances at a price point with sufficient market demand. A final purpose of these collaborations was strengthening mCHP's position in the competition with other technologies, such as heat pumps. The following excerpt from an interview illustrates this:

This means that we need to show the competition which has competing products, for example heat pumps, that our technology is a good one. And then, once out technology – micro CHP – is established and has reached a certain market penetration, we can start competing against each other once again. (translated from German)

Particularly one interviewee, who was leading many of the efforts to cooperate to promote mCHP, stressed repeatedly that the aim of these efforts was to achieve a fair treatment for mCHP vis-à-vis other technologies whose backers he accused of using unfair practices in some instances to give these technologies an unfair advantage over mCHP or disadvantage mCHP unfairly. Many of the activities outlined in Sect. 5.2 were driven by this motivation for which the following quotes are exemplary:

We don't need a bonus, we only need a fair treatment. And the advantage shouldn't come and isn't from the standard, but the advantage is from the real world and the standard should reflect the real world in a fair way.

I had the suspicion that they wanted to get a privileged position of, for instance, electrical heat pumps by pushing micro CHP down.

5.1.3.1 Industry Actors Not Supporting mCHP

Despite these observations of broad collaboration in the heating industry to drive mCHP forward, this did not concern the entire industry. One major appliance manufacturer with little involvement in mCHP technology was critical about these efforts. Representatives of this company participated in standardisation committees and working groups at the industry associations in order to prevent what they saw as formulating rules which would give mCHP an unfair advantage over other technologies. An interviewee working for this company relayed the opposite narrative to that of the supporters of mCHP, claiming that their activities were geared towards giving mCHP unfair advantages over other technologies:

> I am not a friend of the manner how one tried this [Stirling-based] appliance with the corresponding label[4] – because all of this no longer has anything to do with physics. This is just about marketing. And in this place – I know we also have to sell our products – but we as [company name] still try it in a reasonably fair way and this is not fair anymore. (translated from German)

The interviewee voiced his admiration for what he saw as one company with particularly strong interests in the technology pulling an entire industry on their side. He claimed to also speak on behalf of other companies that were sceptical about the rest of the industry's efforts but which were too small to effectively participate in the activities related to standardisation and regulation. This difference in viewpoints about mCHP technology and the cooperation in the industry then led to major conflicts during the development of standards and regulation (see Sect. 5.2.2).

5.1.4   The Role of Intellectual Property in the Industry's Collaboration

Based on our literature review, we expected IPRs to play an important role in the collaboration between different actors in developing mCHP. In particular, we assumed that they would be important in standardisation

---

[4] See Sect. 5.2.2 for details regarding this issue.

for mCHP. We therefore specifically asked interviewees how they had dealt with IPR as part of their NPD and standardisation activities.

5.1.4.1 Protecting Intellectual Property Related to mCHP Technology

The interviews show that IPR was indeed an issue that they considered and that they aimed to protect their innovations where possible. Based on these observations, the interviewed companies can be divided into (1) two companies which considered IPR an important strategic issue and (2) a larger group where IPR was dealt with as a lower-level issue.

Two of the interviewed smaller start-ups stressed that it had been essential for them to think about IPR strategically while building their business. One of them was initially launched with the aim of building entire mCHP appliances but later focused on supplying advanced fuel cells to others in the industry. In this role, keeping the IPR of the fuel cell designs and either producing them on behalf of the customers or licensing the designs was key to the company's business model. The other company in this group also carefully considered how to best use IPR protection to support their business, as the following quote shows:

> We talked about the GSE board, the burner control and the essential air sensor where we place great importance on having the [intellectual] property ourselves. We therefore have patents. We are interested in the Hot BOP, Hot Balance of Plant, we wanted the stack ourselves. There we wanted to have ownership. In this area, in coatings, in compositions and the burner itself, we have patents. We want to be the owner of key parts. But otherwise – and this is part of our strategy, also to keep costs down in this area – we developed the relevant parts together with our suppliers. We have often done this and then afterwards made the part available to our competitors or other actors in the market. (translated from German)

The larger part of the interviewed companies, including the large established players, treated the IPR issue in a more matter-of-fact way. They saw the topic as one that needed to be taken into account when managing mCHP's development but did not portray it as a topic with strategic relevance similar to how this was seen by the first group. The following quote illustrates this approach:

> In some parts we built [intellectual property] ourselves and applied [for patents] ourselves. And we naturally conducted patent searches. This is

even more important, to make sure that you do not introduce something as a product which you may not introduce, quasi conducting a patent violation with the product. This is something which belongs to a product development process by default. The patent search about what one wants to introduce, what one wants to develop. This is an item in the product development process. (translated from German)

5.1.4.2 (Not) Using IPRs in Standardisation for mCHP

While interviewees recognised the importance of IPR in developing mCHP in general, they did not consider the topic as relevant for standardisation. Indeed, when asked about how IPR issues were addressed in the standardisation process, interviewees saw no link whatsoever between the two topics and sometimes were even surprised that such a link was suggested. They claimed that practices such as declaring patents as standard-essential and basing standards on an individual party's IP have not been used in the mCHP context and even were unheard of in the European heating industry, as the following excerpt from an interview shows:

> *Interviewee 1*: There was no such thing [attempts to place IP in standards] here, no.
> *Interviewer*: Okay, this means that this is not common in your industry?
> *Interviewee 2*: No. In any case not in the context of standards. Of course, obviously one tries to protect one's intellectual property, maybe also if one sees that one can trigger something at the competitor. But especially in the fuel cell area and standardisation, or CHP and standardisation, this was not a big topic. (translated from German)

Beyond this, the interviewees even considered bringing IPR issues into the standardisation debate as counterproductive and as being contradictory to the purpose of standardisation. They shared an approach to standardisation which strived to write standards that support all companies in designing their own mCHP appliances, rather than applying solutions that were covered by one party's IPRs. Interviewees also argued that it would not be in their own long-term interest to place their IP in the standard, thereby limiting other companies' options in developing their technological approaches for mCHP, because this would weaken the development and eventual chances of market acceptance of the technology as a whole. The following two excerpts from interviews exemplify these arguments:

Interviewee 1: No. Patents can actually not play a role in standardisation. At least, I have no examples in our area. (...)

Interviewee 2: (...) If you have developed something technologically and you think that you should protect this for yourself, then you register this [as a patent]. But if you want to develop this into a standard, then you initiate a standardisation committee (...) so that you eventually get a standard which you can build into the product and sell without hindrance or [also decide to] leave out [of the product]. (translated from German)

Interviewee 1: We have of course tried to place our own ideas in the standards without revealing, for example, what our safety concept looked like. Especially in early phases, we tried not to show in too much detail what we were doing, especially for the safety concept. And there one always has to achieve a balance.

Interviewee 2: So, enabling the own concept without revealing it and recognising the same at the colleagues from our competitors and leaving them the same wiggle room. We had no interest in preventing or hindering competition in this early stage because this would have weakened the technology as a whole. (translated from German)

The reason why such an approach was seen as weakening the innovation was that it might have caused other actors in the industry to lose interest in mCHP. Following on from the reasoning for collaborating across the industry (see Sect. 5.1.1), this was seen as a potential problem because it would have left the company alone in promoting the technology, e.g. in discussions with government, which would have been unlikely to succeed:

It would have been an extreme risk to weaken the technology in this way and suddenly being left as the only vendor, which would definitely not have been constructive. If the entire [German industry association] had not been interested, [company name] could also not have gone to Berlin on its own to accomplish anything there. Because of this, the others, the competitors had to remain interested in the whole thing. (translated from German)

5.1.4.3 The Overall Impact of IPR on mCHP's Development

Overall, IPRs were considered an important element of managing mCHP's development by the industry. We observed broad consensus among interviewees that protecting own technological developments was important, also when cooperating with other parties. However, there was equally broad consensus among interviewees that IP had no place in the development of standards for mCHP. The interviewees who spoke on this topic all agreed that including proprietary knowledge in the standard would have been counterproductive and eventually resulted in substantial difficulties for the technology's development and eventual success.

5.2   Conflicting Interests in Standardisation and Regulation for mCHP

As outlined in Chapter 3, several standards needed to be changed or newly developed in order for mCHP to be sold into the European market with the intended value proposition. On most questions, such as electrical installations in buildings, other players in standardisation committees adopted a constructive approach towards the innovation. With their support, standards were adapted so that they would accommodate mCHP and provide a basis for the technology's safe and efficient operation. However, two areas of standardisation turned out to be controversial because of competing interests by actors from other technological fields: (1) Questions related to connecting to the electricity grid and (2) developing a calculation method for mCHP's energy efficiency based on the European Union's requirements for energy labels (part of the product standard EN 50465). In addition, several interviewees identified reuse, recyclability, and reparability (RRR) as a new field of standardisation with relevance for mCHP where they expect potential conflicting interests in the future:

> According to a new mandate, RRR – meaning reuse, recyclability and reparability requirements – must also be included in the standard. What exactly this contains is now under discussion. (translated from German)

Because the questions related to the electricity grid and the efficiency calculation method are recurring themes across our interviews and many interviewees stressed their importance for the development of mCHP, we focus our discussion of standards' and regulation's evolution on these two areas.

5.2.1 Standards and Regulation for Connecting to the Electricity Grid

As outlined earlier, being able to connect mCHP appliances to the electricity grid and feeding the generated power into the grid were key to implement the innovation's value proposition. This key importance made the topic one of the focus areas in the standardisation and regulation efforts. During this engagement, the actors from the heating industry encountered a range of stakeholders from other industries, most importantly the electricity grid operators, who were used to a different approach to standardisation:

> There are various actors, typically settled in the energy business, or around the energy business. And for them [the actors from the heating industry], these are quite uncharted waters although meanwhile they have been acting more and more confidently. (translated from German)

> Feeding into the electricity grid is usually shaped monopolistically because utility companies typically used to have monopoly structures. (...) They were not used to developing standards in the same way as, for example, in the gas or (...) household appliance industries, where notified bodies, manufacturers and users sit together in standardisation committees and are looking for compromises. For feeding into the grid, this is different. It has been a long process and we have not yet arrived at the goal that there is equal representation in committees (...). There [in this field of standardisation], one is used to the grid operators determining what [rules] apply. (translated from German)

In the remainder of this chapter, we describe the industry's efforts in dealing with the opposing interests in this field. We start by outlining the environment in which the industry found itself and the conflicting and converging interests resulting from this. We then explain how the stakeholders interacted and how the conflicts between them were eventually resolved.

5.2.1.1 Background: Electricity Grid in Transition

At the time when mCHP's developers worked on the topic, several parallel developments occurred, such as the spread of renewable energy sources and the exit from nuclear power in Germany. These developments had (sometimes substantial) implications for the electricity grid. Traditionally the electricity grid was built around a small number of large power stations, meaning that electricity production could be relatively easily balanced with demand for electricity. With the new developments, a large number of small electricity producing appliances (including mCHP appliances, solar panels, wind turbines, etc.) started appearing in the grid which resulted in substantial changes to the grid's structure:

> Around 20 years ago, we had maybe, say, 1000 generators in Germany and now we have 20 million or 15 million or some number in that range, if you include all the solar panels that feed into the grid. (translated from German)

Furthermore, the spread of renewable energy also means that parts of the electricity production can no longer be adjusted to demand fluctuations because it depends on factors like sunshine and wind. This made mCHP one of several factors[5] in a major transition, which challenged grid operators' and utility firms' traditional approach to managing the electricity grid. According to most interviewees, mCHP was therefore met with certain degrees of resistance by some of these actors, while others participated in partnerships to develop the technology (see below).

> If you look at what the four big [German utility companies] have lost in market capitalisation through shutting down nuclear power stations, through the increase in photovoltaic, through the prioritisation of renewables before [other energy sources], and the fact that for economic reasons the most modern gas fired power stations are not operated anymore today, even though they would produce the lowest emissions out of the fossil [fuels]. And then, politics exerted such a massive influence on the industry that they [grid operators and utility companies] fight helping any other sector tooth and nail. They have so many problems of their own (...) and that's why they resist helping even the smallest CHP or even developing understanding. If you want to see it positively, it is slowly beginning [to change], but much too slowly. (translated from German)

Given this background, some interviewees reported that the established players in the grid field sometimes made demands based on their experience with large power stations, which the interviewees interpreted as

---

[5]Although in the grand scheme of things, mCHP was a comparatively small factor relative to the other developments.

aiming to hinder mCHP's development by imposing unreasonable requirements in the standards and regulation:

> *Interviewee 1*: In standardisation and regulation on the electrical side (...), they crack nuts with sledgehammers and we often came across attempts to prevent technology through standardisation.
>
> *Interviewee 2*: They really put obstacles in one's way. I am thinking of one example regarding how the amount of electricity that is produced by an mCHP appliance should be measured and where the measurement device should be placed. Traditionally, it is clear that, if you build large equipment, then you have some (...) measurement device (...) and if this is not directly on the turbine it is in an electrical cabinet far away. And one tried to transfer this concept to a small electricity generator [even though there] you do <u>not</u> have a separate electrical cabinet (...) but everything that is needed for the operation has to be built into the appliance, into one enclosure. (translated from German)
>
> On the grid connection side we had the occasional discussion because the utility companies inherently have a different view on the technology. I remember a discussion (...) where the utility companies (...) wanted to draw upon a standard to enable communication between the fuel cell and a higher-level control unit to create a 'virtual power station' (...) and where we said 'wow, that's totally excessive, they want to impose a standard on us that can communicate with a network control centre and that would ask way too much from our appliance'. (translated from German)

5.2.1.2 Converging and Competing Interests with Other Technologies

As the development of mCHP coincided with other technologies' emergence, the actors in the heating industry were not only confronted with the traditional grid operators and utility firms, but also with the interests of these other technologies' developers. Most importantly, the needs of renewable energy sources (which also enjoyed some political support) were a major factor in the development of standards and regulation for grid access. In some cases, the heating industry's interests converged with the ones of these other actors. For example, mCHP was seen as a potential technical solution to ensure grid stability in the future when renewable energy would make up a large part of the electricity generating capacity, thus providing complementary value:

The idea is basically that one can smoothen the volatile energy production of renewables a little bit with a large number of mCHP appliances in the grid. Because when you look at the energy generation curve of an mCHP appliance, this is quite complementary to a photovoltaic module. (...) When the sun is shining heavily, I don't need heat and the mCHP appliance does nothing. When a lot of heat is required – usually in the winter, in the evening, or in the morning – then I have electricity generation from the mCHP appliance. (translated from German)

The interests of mCHP's developers and other technologies' proponents conflict ed on other questions. One example that was mentioned in several interviews is the requirements for dealing with frequency changes outlined in Sect. 3.4.2, which poses a substantial hurdle for Stirling-based mCHP appliances. The introduction of this requirement was driven by the expectation that large sudden changes in wind or sunshine would make the grid frequency volatile when many renewable energy electricity generators are connected.

5.2.1.3 Activities in Standardisation and Regulation for the Electricity Grid

Given this background of an electricity grid in transition and other technologies developing in parallel, the interviewed actors aimed to influence standards and regulation so that workable solutions for mCHP could be found. Our interviewee at the European industry association summarised this goal as follows:

To be able to feed the one kilowatt [of an mCHP appliance] into the grid, the supporting conditions must be right. There must not only be supporting conditions for 500 kilowatt [appliances]. This is like traffic on the roads. If you have lots of racing cars on the roads, they of course have other interests, they drive at different speeds than (...) a small car in between which can only drive 100 instead of 250. (...) And therefore, a compromise has to be found where we say 'he may also use the road, but he may only drive in the right hand lane'. (translated from German)

To reach this goal, the actors engaged in standardisation and regulation pursued various activities to increase the impact of this engagement. These activities can be grouped as (1) forming coalitions, (2) establishing evidence about the technology and informing other stakeholders about its needs, and (3) adapting mCHP technology itself where necessary and possible.

The first group of activities (coalition forming) was in many cases based on the collaboration forums outlined in Sect. 5.1. For example, the 'Callux' project that was undertaken as part of the IBZ in Germany included several energy suppliers as collaboration partners. Especially smaller, local energy suppliers sometimes saw mCHP as an opportunity to shift the balance of power generation away from centralised power stations owned by their large competitors. Gas suppliers who *"were interested in selling gas"* (translated from German) were also supportive of mCHP in questions related to grid access. However, being able to form these coalitions and operate these field trials was not always easy, as the following quote shows:

> It already started with having to find people who conducted field trials together with us. Of course, these appliances then also have to be approved, that is clear. But these were people who, let's say, accommodated us with a certain goodwill and then maybe also interpreted grid connection rules generously and did not make it impossible from the start. Because they knew that these were small appliances with initially small quantities. (...) [And these people] also saw new business opportunities in the technology [although] it took a while for the utility companies to recognise these opportunities. (translated from German)

Such collaborations across stakeholders also were directly linked to informing stakeholders, making them aware of the technology, and establishing evidence about it. This second group of activities was necessary because many actors involved in developing requirements for grid access were unaware of the technological characteristics of mCHP:

> But they [the grid operators] of course have their large power stations and rotating machines with their inertia in mind. Feeding into the grid with a small appliance – the needs that exist there were not in their focus. And there we needed to vehemently [argue] on the European level when the Network Code Requirements for Generators [were developed]. (...) And it was not easy to convince these circles that mCHP behaves in a special way. When you switch an mCHP appliance off, you need to restart the thermic process. But they assume that the rotating machine runs anyway or that a solar panel can immediately feed electricity into the grid when you switch the semiconductor. (...) A fuel cell needs to be restarted. This takes minutes and they want to switch it on immediately at the right

frequency. These are basic principles which are difficult to convey. (translated from German)

Neither had we experience with the electricity generating sector, nor did the electricity generating sector know anything about these small generating appliances. And only once the electricity producers realised that these small generating appliances must be taken seriously, that they are not a temporary phenomenon (...) [but] actually enter the market, then one also reacted accordingly in that group, respectively started trying to establish the rules. (translated from German)

To support this information of other stakeholders, the developers of mCHP relied on evidence created by field trials, such as the 'Callux' project mentioned above where *"a few hundred fuel cell mCHP appliances were brought into the field"* (translated from German) and their effects on the electricity grid were measured on behalf of utility companies by an independent research institute.

Finally, the developers of mCHP also adapted their technology to make it more acceptable to other stakeholders in the electricity grid. Some interviewees stressed that the interaction with these stakeholders helped their understanding of the issues faced by the electricity grid operators and mCHP's possible positive and negative impacts. This increased awareness allowed them to facilitate these other stakeholders' concerns and sometimes even work out technical solutions jointly with these actors, as the following quote shows:

For example, there was the need to cover wider scopes of grid frequency and different technical solutions existed for this [issue]. And the one which we preferred and also finally implemented (...) [was based on] considerations which we worked out together with the grid operators and the power station operators in this VDE [Verband der Elektrotechnik, German association for electrotechnology] committee. (...) [And there would have been other solutions which] would not have been so accommodating for us, which would have been much more expensive. (translated from German)

5.2.1.4 Limited Influence on Standards and Regulation for the Electricity Grid

Despite the efforts to influence the development of standards and regulation, the actors in the heating industry remained relatively small players in the field with limited influence on the process. Some interviewees acknowledged this as a problem for dealing with issues related to these requirements:

> *Interviewee 1*: The difference to the standards that we talked about a moment ago [standards relating to gas-safety and efficiency] is that we get the standards [relating to the electricity grid] on the table and we have very, very little influence to make a difference there.
> *Interviewee 2*: The electrical side is extremely difficult.
> *Interviewee 1*: Exactly. There are also completely different structures and [company name] is not necessarily a big player – I would even say – not at all. (translated from German)

Consequently, the actors in the heating industry were not entirely successful in reaching their goals. The rules for dealing with grid frequency changes mentioned in Sect. 3.4.2 are an example where the heating industry's limited influence on the process made it unable to prevent a change in the standard that was against their interests. These rules were introduced during the development of mCHP, replacing earlier requirements that were easy to fulfil for Stirling-based mCHP appliances:

> The requirements for connecting to the grid. (...) There was a standard and we complied with that standard and then what was previously required was now forbidden or the other way around. So there, the standards are not fixed situations, they are temporary.

Technical solutions to design Stirling-based mCHP appliances in line with these changed requirements have a high impact on the devices' costs and efficiency. At the time when we conducted our interviews, the companies using Stirling engines relied on provisions in the grid access regulation which exempt new, innovative technologies from certain requirements and allow them to continue operating according to the old requirements (see European Commission, 2016, secs. 66–70). However, these temporary provisions only apply until a limited number of appliances using the new technology have been connected to the electricity grid. Consequently, the actors relying on Stirling technology were still in the process of working on this issue at the time of our interviews:

> We've been fighting that [the new requirements] for two years and there's hopefully a special dispensation within that.

5.2.2   Conflicts Surrounding the Calculation Method for mCHP Appliances' Energy Labels

A second major topic of standardisation was the calculation method for assessing mCHP appliances' energy efficiency, which underlies the efficiency label that each appliance needs to carry according to the ErP and Energy Labelling Directives (see European Parliament & Council of the European Union, 2009, 2010). The topic was particularly important and contentious due to its relevance for European legislation and the European Commission's involvement in the standardisation process.

The calculation method is part of the product standard (EN 50465, the latest version of which was published in 2015), which did not yet exist when the technology's development started (see Sect. 3.1).[6] This standard "specifies the requirements and test methods for the construction, safety, fitness of purpose, rational use of energy and the marking of micro Combined Heat and Power appliance[s]" (CENELEC, 2017). While development of most of the standard's elements proceeded relatively smoothly, there were major conflicts regarding the energy efficiency calculation methods:

> Within standardisation, the range of opinions about calculating the efficiency was, in my opinion, the biggest problem. (translated from German)

These conflicts related to two fundamental issues: (1) There was disagreement about the formula which underlies the calculation and for which different options were being discussed. (2) The way in which the European Commission was involved in the process was seen by most actors as exceeding the role that it should play in developing harmonised standards (also see the explanation of harmonised standards in Sect. 3.2.1).

Actors from the heating industry were the major players when developing EN 50465. Because this standard only covers mCHP appliances, parties who had high stakes in the technology (mostly overlapping with the actors covered in Sect. 5.1) dominated the relevant committees where it was developed. In addition, European consumer

---

[6]EN 50465 was an already existing standard on gas-powered fuel cells which was extended in scope to cover all mCHP appliances, rather than developing an entirely new standard to fill this gap.

and environment protection NGOs were involved although, according to the interviewees' depiction of the process, these actors did not have a major impact on the outcomes. The European Commission was not represented in the committees but nevertheless influenced the standard's development in a major way.

Below, we first outline the conflicting positions regarding the calculation method. We then summarise the conflicts between the heating industry and the European Commission during the development process. The chapter then ends by describing the process's outcome and giving an outlook to future developments expected by our interviewees.

5.2.2.1 *Conflicting Positions Regarding the Calculation Method*

Deriving a calculation method to assess mCHP appliances' efficiency was not trivial because this formula needed to incorporate both the heat and electricity produced by mCHP appliances and at the same time give a result which would allow a meaningful comparison with other heating technologies for consumers:

> And now you have an additional problem: How do you grade this new segment, which delivers two forms of energy as an output, among the existing heat generators and energy products? (translated from German)

Consequently, there were different views regarding how the electricity produced by an mCHP appliance should be rewarded when assessing the appliance's energy efficiency:

> There were companies who wanted to have this calculated in specific ways. We even had three different methods before we finally agreed on one in a compromise [within the industry association]. (translated from German)

Most of the industry agreed on this compromise, which was developed in standardisation committees and industry association's working groups. However, a minority of industry actors including one major appliance manufacturer (also see Sect. 5.1.3) was in favour of a different method, which was also supported by the European Commission. These different preferences for calculation methods resulted from different views on how to consider aspects like the produced electricity, reduced needs for electricity from (relatively inefficient) power stations, and where to draw the boundary of the system for the purpose of assessing its efficiency:

> There were long discussions about where the system boundary of the appliance lies. How do you actually calculate the efficiency of such a Stirling product? Do you include the efficiency of the boiler or do you only take the efficiency [of the Stirling engine]? And finally, we brought ourselves to write into the standard that the entire system is considered. (translated from German)

The parties disagreeing with the industry compromise argued that using this formula is inappropriate for assessing an mCHP appliance and that the underlying approach would only be suitable for assessing the energy efficiency of an entire building but not of a standalone heating appliance. They accused other actors in the industry to push this formula through in order to make their appliances look more energy efficient than they actually are, stating that *"this no longer has anything to do with physics [and] is all about marketing"* (translated from German).

On the other hand, interviewees supporting the industry compromise argued that this was the best way to reflect physical realities and ensure that the results enable consumers to compare mCHP to other technologies. They claimed that the alternative formula did not sufficiently factor in the electricity produced by mCHP appliances in addition to heat.

> And this [the alternative formula] was in such a way that electrical heat pumps were clearly treated preferentially in the resulting efficiency values, compared to micro CHP. And then we intervened and said: 'The micro CHP appliance cannot be nearly put on the same level as classic condensing boilers. And a heat pump has an efficiency value up to a third higher compared to the micro CHP, this is not reasonable.' That a heat pump has a higher efficiency than a classic condensing boiler is clear. (…) This is absolutely OK. But how does an mCHP appliance fit into this? (translated from German)

This view of the alternative calculation method being wrong was also supported by an interviewee at an academic engineering research institute based at a German university:

> One of the colleagues made a nice example calculation. (…) Same primary energy in, (…) identical amount of useful energy out. And then he (…) applied the EU calculation for the labels. And for a heat pump-based solution he got an A++ and for the micro CHP-based solution, he got an A+. This means that the methodology of the European Commission is wrong insofar that two different technologies generate the same useful energy

with the same input of primary energy but get different labels. And there, the working group said: 'No that cannot be the case, this is physically wrong. And it is also confusing the customer.' (translated from German)

5.2.2.2 Interactions Between the Industry and the European Commission

Throughout the standardisation process (including before a formal standardisation request was made to CEN/CENELEC), the European Commission promoted an—in most interviewees' eyes—unjustified calculation. Together with the 'group dynamics' outlined in Sect. 5.1.3, this caused strong resistance among mCHP's developers and also made the topic highly emotional for some of them. In their view, the European Commission had overstepped their role in supporting this contentious formula which they saw as problematic:

> There was a high level of frustration within the standardisation committee because the engineers simply said: 'Hey, we are (...) calculating in the physically correct way. And if anybody can calculate correctly, that is us, the engineers, and not the civil servants. (translated from German)

> It is not so easy for them [the European Commission] to see what their real role is. You see a kind of imperialistic approach. On the one hand, the Commission wants to regulate technical details and technical content which is not according to the New Approach and where they don't see their role. Are they a stakeholder? Are they forcing something? So, I think (...) there's a problem area here.

Initially in the process, the industry faced unclear guidelines from the European Commission:

> At certain moments in that standardisation group we saw [that] we seem to be shooting at a moving target. There was from the side of the Commission and the consultant, which the Commission had appointed, a kind of calculation model which became more complex and more complex and more complex (...). And then, at a certain moment, the Commission changed their ideas about the calculation procedure and then it seemed that we were (...) shooting at a moving target. So then, in the standardisation committee, we said 'we will put this on ice for a certain time, first see where the Commission will move and where the negotiations between the associations and the Commission will move'. And then, finally, we had an agreement with the Commission that we would propose a standard and then we would discuss it. And then we went ahead and took the initiative again.

As the ambiguity of the European Commission's position on this issue eventually ended, it became obvious that the European Commission favoured a different calculation method than the compromise supported by most of the industry (see above). Given this situation, the members of the standardisation committee nominated two representatives (one of our interviewees at an appliance manufacturer and the consultant who accompanied the industry) to negotiate directly with the European Commission (also see Sect. 5.1.2). Both of them described these negotiations as very difficult because the process was lacking transparency from their perspective. They had the impression that other parties' lobbying and political interests not directly connected to mCHP influenced the European Commission's position to a large extent, but it was not transparent to them who was behind this influence and which arguments were used by these parties. Nevertheless, there was a clearly visible bias in favour of renewable energies at the expense of mCHP:

> I have seen many drafts [from the European Commission] of these requirements over the last five years. And in one draft, they had an explanatory memorandum. And there (…) they said: 'Micro CHP is an efficient technology but it is not renewable, it is not solar or wind power (…). And therefore (…) it should come to a result which is lower than renewable.' And then they said 'renewable is defined if the efficiency is at minimum 115%, so the efficiency should be below 115%'. Completely not logical, and it shows indeed that they were very biased.

> And finally, at some point there was a comment from the European Commission – of course only verbally and not in writing – 'we don't need to discuss this anymore, micro CHP ought not be better than A+, full stop.' (translated from German)

The European Commission's support for its preferred calculation method was documented in Commission Communication 2014/C 207/02 (European Commission, 2014).[7] This communication took many actors in the industry by surprise:

> I saw the latest draft which was going to the parliament and then I saw these words and I thought: 'Oh, what now? Now they're choosing already

---

[7] Such a Commission Communication is an official document where the European Commission outlines its policy on a specific topic (Overy, 2016).

although we had the agreement that we would first have a discussion and then be able to exchange arguments etcetera. And now they have done it this way.' So at first instance, it was very disappointing.

Around ten months after publishing the Commission Communication with its preferred calculation method, the European Commission released a formal Standardisation Request on the matter (European Commission, 2015).[8] This request asked industry, among other things, to develop a standard that specifies energy efficiency calculation methods for mCHP. Several interviewees pointed out that this request was released with a tight deadline and "*came when the standard was finished almost*". Furthermore, they mentioned that the earlier events implied that the standard was expected to use the European Commission's calculation method as a foregone conclusion.

While this conflict with the European Commission was ongoing, there were also discussions within the industry about the best way to proceed. As part of this process, some actors sought expert advice about the legal implications of a Commission Communication, which revealed that it was only an opinion of the European Commission and was not legally binding. This encouraged these actors to keep pursuing the compromise found earlier within the industry. However, other actors were in favour of proceeding with the European Commission's formula, as the following exemplary quote from our interview with a representative of the industry association shows:

There were definitely also different opinions [in the industry]. And some also gave up and said: 'No, this is not the way it goes. I am sticking my head in the sand, just do whatever you want.' Again, the standard is [based on industry] consensus and all [industry actors] committed to it. But especially for the efficiency calculation [where] the Commission had different ideas, there also were actors [who said] 'it doesn't matter what our opinion on this is, the Commission wants this and then we do this'. And there were others who said: 'No, we don't do it this way. We got an answer from the Commission which (...) in our opinion is completely wrong. We want it our way.' (...) We had two meetings with heated discussions about which method is more correct. (translated from German)

---

[8] The European Commission uses Standardisation Requests to initiate development of standards needed to support 'essential requirements' in European directives with the intention to harmonise the resulting standards (see Sect. 3.2.1).

Much of this discussion revolved around whether to prioritise the standard's harmonisation or a physically correct calculation of mCHP's energy efficiency. One interviewee highlighted that it was foreseeable that the European Commission would not harmonise a standard with the formula favoured by most of the industry. According to this position, which was shared at the time by the British national mirror committee, it could not be in the interest of anyone in the industry to develop a standard that would eventually not be harmonised by the European Commission. Other interviewees did not see this as a major problem. Because the energy labels are based on self-declaration,[9] appliance manufacturers would be able to choose which formula to base their labels on, even if the standard was not harmonised. In this scenario, it was uncertain whether and how the national market surveillance authorities would react but the majority of the industry considered the risk of negative consequences small. They expected that applying a standard developed by an ESO would give them good arguments in a hypothetical investigation by the market surveillance authorities, even if the standard was not aligned with the European Commission's position.[10] They therefore saw an—in their eyes—fairer calculation method as more important than the standard being harmonised under the ErP and Energy Labelling Directives. In addition, they expected that the product standard could still be harmonised under the Gas Appliance Directive due to its gas-safety-aspects.

At the end of these discussions, the supporters of the European Commission's calculation method were outnumbered and the committee put a draft standard to vote at CEN/CENELEC. This draft included the energy efficiency formula supported by the majority of the industry and was transparent about the issues in the standardisation process. This caused the European Commission to intervene in CEN/CENELEC's voting process, although this intervention was eventually unsuccessful:

---

[9] This means that companies may calculate their products' efficiency themselves and use the appropriate energy label. Notified bodies are not needed for certifying a product's energy efficiency.

[10] One interviewee deviated from this position: In his opinion, especially in the wake of the Volkswagen Diesel scandal, the industry should avoid any semblance of making its own rules in the matter which deviate from regulation. However, the majority of actors in the industry argued that an—in their eyes—physically correct formula was more important, also from these ethical and public opinion points of view.

Finally, we have written a foreword to the standard to make completely transparent – for the people who had to vote on the standard – that the standard was deferring from the Commission Communication, which is an opinion of the Commission without binding effect. And the standard was finally accepted but the Commission several times tried to intervene and really obstruct the voting process. So, they first asked – (...) As joint working groups, as technical committee, we had decided 'we are going for a formal vote'. We sent it to CENELEC for formal vote and first the Commission asked CENELEC not to send it for formal vote but CENELEC did. Then, they asked CENELEC to stop formal vote, even in the middle of the process. And finally, in the last step, after the vote was positive, there was a ratification by the technical board of CENELEC. And they tried to influence the technical board not to ratify the standard. So, in fact, three times they really tried to obstruct the standard and they didn't succeed.

There also was the story that CEN/CENELEC published the standard and the EU Commission reprimanded CEN 'how can you publish something that has nothing to do with our mandate?' Whereupon the top level of CEN got into the game and said: 'Just a moment, slowly. You may give us a mandate but we are completely independent about how we write our standards and what we write in them. Because it is us who have the technical expertise, and you don't.' There was a quite interesting exchange of letters between the Commission and CEN where the top level of CEN distanced itself and said (...): 'We are writing technical standards. And if our engineers consider this standard correct from a technical point of view, then it is correct from a technical point of view.' (translated from German)

5.2.2.3 Outcome of the Conflicts and Outlook to Future Developments

Looking back at the process, most interviewees remained critical of the European Commission's role. However, two interviewees in particular also reflected critically on the industry's activities. One of these interviewees questioned whether it was wise to accept the European Commission's standardisation request, given the development of the process up to that point:

The problem is that one does not (...) occupy oneself sufficiently with the mandates [before accepting them]. The mandate goes to CEN/CENELEC, goes to the working groups [and] the committees, there is an appeal period when one can say 'this is nonsense, we are not interested'. This did not happen in this case and then, at some point, [the mandate] is accepted. And then it is on the table and one is stuffed. (translated from German)

The second interviewee concluded that involving additional stakeholders in the process might have been helpful in addressing the issues with the European Commission:

This clearly is something that did not go well. Maybe, we would have had to involve the national governments much stronger? Because the Commission is not deciding on its own and it is always easy to say 'yes, the European Commission (...), that circle does not appreciate our course of action'. But if we had activated the country representatives of different countries at an early stage, for example [commissioner] Oettinger in our case... (translated from German)

Nevertheless, EN 50465 was eventually published including the calculation method favoured by most of the industry. As foreseen during the standardisation process, this meant that the European Commission did not harmonise the standard under the ErP and Energy Labelling Directives. When the standard was published, the UK mirror committee included a national foreword in line with its earlier position in the British version of the standard, advising against the use of the calculation method included in the standard:

The UK committee advises, for the calculation of μs and μson of cogeneration space heaters the methodology described in the Commission Communication, reference 2014/C 207/02 should be used. This method is robust, scientific, provides a fair comparison across all technologies and is aligned with the established methods for assessing and comparing cogeneration performance. (BSI, 2015)[11]

---

[11] Clearly, the foreword to the standard was written before the Brexit referendum... Nevertheless, some interviewees also found this remarkable:

Interviewee: As I already said, as often in Europe, the Brits think that they need to do their own thing. And they do this thoroughly.

Interviewer: (Laughing) Only this time with the unique situation that they share an opinion with the European Commission.

Interviewee: Yes, in this case they agree with the European Commission. This really is – one should make a big poster of this and put it up on the wall somewhere. Happens seldom enough... (translated from German)

Despite this standard not being harmonised, most companies in the industry have so far been using it in calculating their appliances' energy efficiency for the self-declared energy label without negative consequences from the national market surveillance authorities:

> The Ecodesign and the Energy Labelling Legislation have started to be applicable from September 2015, so that is two years ago now. And I think (...) the vast majority of companies have been using the standard and also the calculation method of the standard. I know of one exception which is using the Commission Communication and the regulation and which really, I think, is using it to their own advantage.

In our final interview in August 2017, we also learned that the European Commission has in the meantime started its regular review of the directives in question. As part of this review, the Commission also ordered an assessment of the directives' impacts:

> *Interviewee*: Currently, the process of review of the legislation is starting, or has started some months ago. The European Commission has already announced that to us as CHP representatives. Now the regulation is written but then you have new chances. They had their attempt to change physics but they were open for review and improvements of the legislation during that official review, which was announced that it should be ready, I think, five years after adoption of the regulation. (...) At least, they have ordered a consultant to make an evaluation. (...)
>
> *Interviewer*: And then, potentially it could be harmonised after the review changes this legislation?
>
> *Interviewee*: Yes, perhaps. Or, perhaps, the legislation will even be changed more so that the other standards have to follow anyhow.

Depending on this assessment's outcome, the European Commission may therefore change its position on the calculation formula. In addition, fundamental changes to the directives are also possible, if the review finds that they need to be improved. This outcome would possibly also require the industry to develop entirely different standards. The future development of this issue is therefore still open.

5.3   Interviewees' Evaluation of the mCHP Case

In Chapter 3, we presented the various ways in which standards and regulation influenced the development of mCHP, which triggered the extensive company- and industry-level activities depicted in Chapter 4, Sects. 5.1 and 5.2. We also asked every interviewee to evaluate the effects of these activities on mCHP and the relevant standards and regulation. Because all mCHP appliances must fulfil the same set of requirements, these evaluations were similar across manufacturers despite the sometimes-different approaches to managing standards and regulation.

Most applicable standards and regulation were already available and supported mCHP's development before the industry actors initiated their activities (see Chapter 3). These activities therefore mainly focussed on topics where standards and regulation were still missing and/or not supporting mCHP. Because of these efforts, standards and regulation now support mCHP technology in three additional ways: (1) The new requirements for access to the electricity grid provide a workable solution to connect mCHP appliances to the grid. (2) The new product standard defines requirements for safety, energy efficiency, and related topics for mCHP, which support conformity assessment of the technology. (3) Despite the conflicts with the European Commission detailed in Sect. 5.2, the energy efficiency calculation methods in the product standard support the industry in fulfilling the requirements of the European directives related to energy efficiency. Furthermore, some interviewees also mentioned supporting effects of these new standards beyond now being able to fulfil regulatory requirements. They also help the companies in the field to communicate the technology's benefits to their customers and provide confidence to adopters of the innovation.

These changes in standards and regulation enabled the industry to market mCHP appliances in Europe. All interviewees at major manufacturers stressed the importance of aligning their company-level management with the industry-level work to reach this outcome, estimating that they might even not have been able to sell mCHP products at all in the European market without the activities at both levels:

> *Interviewer:* Can you already estimate whether this collaboration between new product development and standardisation was successful or not? Or is the result still pending?
> *Interviewee 1:* This is positive.

Interviewee 2: Yes.

Interviewee 1: It definitely is. We can say that we most likely would not have a product if one had not intensively worked on this. This is definitely very, very crucial, also specifically the network connection requirements (...). It could absolutely have been the case, if we had not worked on this topic and had not been interested in it, that we would not have had a product at some stage. Or a product that does not conform to these standards.

Interviewee 2: This could have happened, yes.

Interviewer: OK, this means that the worst-case-scenario would be that you could not sell it?

Interviewee 2: Yes, exactly.

Interviewee 1: Exactly, exactly. (translated from German)

Consequently, apart from one company which favoured other technologies in its product portfolio, the interviewed major appliance manufacturers have mCHP appliances in the market at the time of writing. While some companies exited the development of Stirling-based mCHP appliances (see Sect. 2.2.2), this was due to reasons unrelated to standards and regulation.

Although the smaller companies did not participate in the industry-level activities to develop standards and regulation, they still benefitted from the changes that resulted from these activities. While the interviewed start-ups did not yet produce mCHP appliances at full commercial scale when we interviewed them, they were confident that their products could be marketed under the partly revised requirements from standards and regulation:

Last year, we reached a milestone which was important for us. We received the CE batch approval for the system. This means that we can install the system in limited numbers across Europe. The next step, which we are taking in parallel to the system's market introduction, is that we seek the full CE mark. This means that we can build an unlimited number of appliances but on the other hand we may then change <u>nothing</u> on the appliance [without having to re-certify it]. (translated from German)

As I already said, we are now at the stage of commercialising [where] it [the appliance] goes to the first customers and the first field tests [and] once it goes out, everything will be 100 per cent adapted to the standards. (translated from German)

In line with these results, the interviewees generally were very happy with the outcomes of their activities but had reservations about the needed steps to get there, as the following quote summarises:

> I'm happy with the results [of the process], I'm not often happy with what we needed to do to get these results. Sometimes, it was really tough and time-consuming, and involving a lot of lobby work and convincing people etcetera. It would have been nice if that had been more efficient.

REFERENCES

BSI. (2015). BS EN 50465:2015. Retrieved December 22, 2017, from https://shop.bsigroup.com/ProductDetail?pid=000000000030249912.

CENELEC. (2017). Project: EN 50465:2015. Retrieved December 12, 2017, from https://www.cenelec.eu/dyn/www/f?p=104:110:542824611155801::::FSP_ORG_ID,FSP_PROJECT,FSP_LANG_ID:407783,37230,25.

European Commission. (2014, July 3). Commission y/C 207/02. *Official Journal of the European Union, C 207*, 2–21. Retrieved from http://eur-lex.europa.eu/legal-content/EN/TXT/PDF/?uri=CELEX:52014XC0703(01)&from=DE.

European Commission. (2015). Commission implementing decision of 27.4.2015 on a standardisation request to the European standardisation organisations pursuant to Article 10(1) of Regulation (EU) No. 1025/2012 of the European Parliament and of the Council in support of implementation. Retrieved December 21, 2017, from http://ec.europa.eu/growth/tools-databases/mandates/index.cfm?fuseaction=select_attachments.download&doc_id=1584, http://ec.europa.eu/growth/tools-databases/mandates/index.cfm?fuseaction=search.detail&id=555.

European Commission. (2016). Commission Regulation (EU) 2016/631 of 14 April 2016 establishing a network code on requirements for grid connection of generators (Text with EEA relevance). Retrieved December 19, 2017, from http://eur-lex.europa.eu/legal-content/EN/TXT/?uri=OJ:JOL_2016_112_R_0001#d1e7879-1-1.

European Parliament, & Council of the European Union. (2009, October 31). Directive 2009/125/EC of the European Parliament and of the Council establishing a framework for the setting of ecodesign requirements for energy-related products. *Official Journal of the European Union, L285*, 10–35. Retrieved from http://eur-lex.europa.eu/legal-content/EN/TXT/PDF/?uri=CELEX:32009L0125&from=EN.

European Parliament, & Council of the European Union. (2010, June 18). Directive 2010/30/EU of the European Parliament and of the Council on the indication by labelling and standard product information of the consumption of energy and other resources by energy-related products. *Official Journal of the European Union*, *L153*, 1–12. Retrieved from http://eur-lex.europa.eu/legal-content/EN/TXT/PDF/?uri=CELEX:32010L0030&from=EN.

Initiative Brennstoffzelle. (2017). *Wer ist die IBZ?* Retrieved November 2, 2017, from http://www.ibz-info.de/wer-ist-die-ibz.html.

Overy, P. (2016). *Update: European Union: A guide to tracing working documents.* Retrieved December 25, 2017, from http://www.nyulawglobal.org/globalex/European_Union_Travaux_Preparatoires1.html#_4._Working_Documents.

CHAPTER 6

Building a Grounded Theory on Managing Standards in Innovation Contexts

Abstract This chapter combines the patterns identified in the earlier chapters into a generalisable grounded theory and identifies the relationships between them. This grounded theory is based on a framework of three nested levels: (1) the company, which is part of (2) an industry, which is in turn part of (3) its wider context. The theory focuses on supporting factors and activities needed on the company- and industry levels to facilitate effective management of standards and regulation in innovation contexts. This chapter also shows how the three levels are linked together. The grounded theory explains how innovators can deal with demands and influences from the wider context by engaging in industry-level collaboration.

Keywords Innovation management · New product development
Cross-company collaboration · Co-opetition · Managing standards and
regulation · Managing societal needs

The empirical insights presented in the earlier chapters provide an excellent base for building theory on our research question and allows us to address the theoretical gaps outlined in Sect. 1.2.4. To do so, we develop a process model of the management. This model includes the activities needed to successfully introduce an innovative product to a regulated market where standards are needed, and a number of underlying structural elements that enable these activities.

© The Author(s) 2019
P. M. Wiegmann, *Managing Innovation and Standards*,
https://doi.org/10.1007/978-3-030-01532-9_6

As we already expected in Sect. 1.2, these activities occur at different levels. Figure 6.1 shows our general framework of the three nested relevant levels. In this framework, (1) a wider context encompasses (2) several industries, which in turn are made up of (3) a number of companies. Concerted activities on all three levels are necessary to align innovation and standards/regulation as achieved in the mCHP case (see Sect. 5.3).

Our further theorising fills in the blanks of Fig. 6.1 by looking closely at each level and identifying the factors which eventually lead to such an outcome. We build detailed theory about the company level (Sect. 6.1) and the industry level (Sect. 6.2). Finally, we consider how all of this relates to developments and the associated processes that occur in the wider context of an innovation (Sect. 6.3). Following these theory-building efforts, we end the chapter with some final thoughts on our findings (Sect. 6.4).

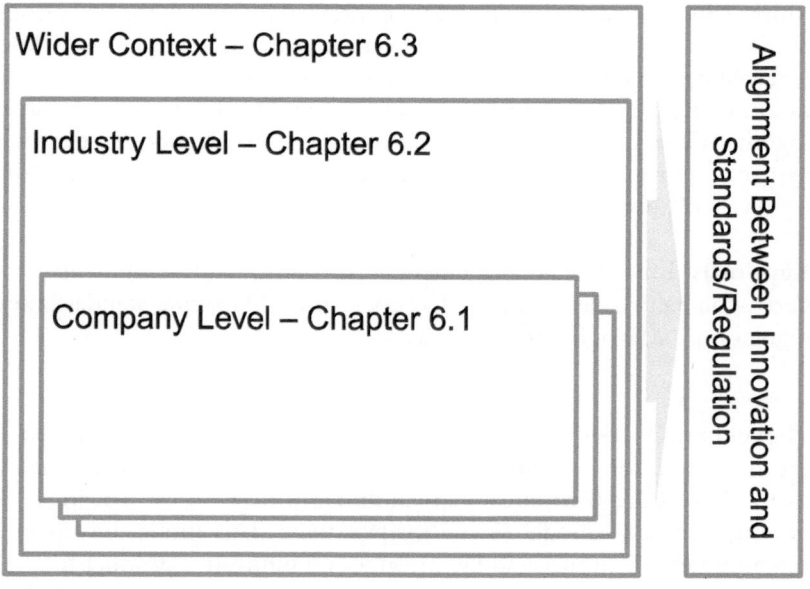

Fig. 6.1   Framework for a theory on managing standards in innovation contexts

6.1   Managing Standards and Regulation on the Company Level

The different types of standards'—and by extension also regulation's—strong implications for innovations make them key issues to manage in NPD contexts. We first consider the company level. In general, the observations from our case show that a number of *supporting factors* need to be in place as necessary conditions to form the foundation for managing standards and regulation successfully (shown in the bottom half of Fig. 6.2 and discussed in Sect. 6.1.1). Building on this, companies need to carry out several activities to ensure that an innovation fulfils all standard- and regulation-related requirements (shown in the top half of Fig. 6.2 and discussed in Sect. 6.1.2). These activities ultimately determine the degrees of freedom for the innovation, as we show in Sect. 6.1.3.

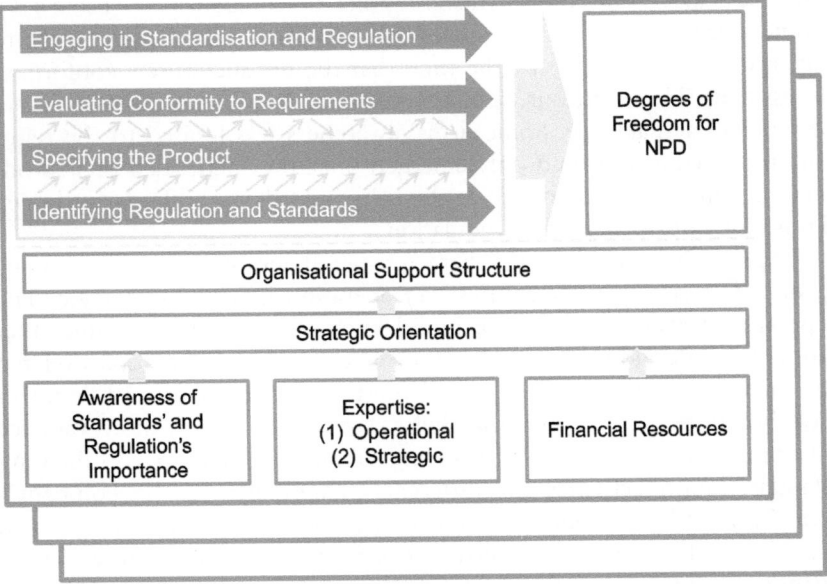

Fig. 6.2   Company-level management of standards and regulation in NPD contexts

6.1.1   Supporting Factors: Necessary Conditions for Managing Standards and Regulation

We observed a number of recurring themes across the interviews (see the data presented in Sect. 4.1), which form the foundation for companies' activities. Having such a foundation in place appears to be a precondition for successfully addressing standards and regulation. On the most fundamental level, companies exhibit three key characteristics (awareness of standards' and regulation's importance, expertise, and availability of financial resources). These three key attributes drive the degree to which the company adopts a strategic orientation which in turn influences the organisational support structure for managing standards and regulation. We provide more detail about each of these aspects below.

6.1.1.1   Key Characteristics: Awareness, Expertise, Financial Resources

Awareness of standards' and regulation's importance is the first key characteristic of companies that our data shows to be relevant. Our interviews demonstrate that companies differ substantially on this aspect (also see Sect. 4.1.1 and the characterisations of companies in Table 4.1). Some degree of awareness about this topic's importance is likely to emerge in any company by the time that the product enters conformity assessment. However, our case shows substantial variation in how aware companies actually are. Some firms' awareness was limited to the regulation-related aspects and only emerged once they addressed their product's certification. Companies at the other end of the scale showed deep knowledge of standards and regulation.

Expertise is a second key characteristic: Relevant knowledge can be grouped in two main categories: (1) operational, and (2) strategic. The operational expertise covers technical knowhow (which companies that are able to develop an innovation are likely to have) and topics related to effective participation in standardisation committees and industry collaborations (e.g. negotiating skills). We observe much more variance in companies' strategic expertise (e.g. abilities related to coordinating standardisation activities for different technologies in the company's portfolio, and contributing to the industry-level processes discussed in Sect. 6.2). This strategic expertise is needed for assessing the effects of standards and regulation and effectively managing the company's input in standardisation.

While much of this expertise is company-internal, all interviewed companies also relied on external expertise in areas where their knowledge was insufficient (in our case mainly coming from consultants and notified bodies). This observation suggests that being aware of the limitations of one's own expertise and seeking outside help where needed is important for successfully managing standards and regulation for innovation. It also suggests that a company's ability to manage these topics relies to some extent on the industry structure, and in particular the supporting institutions (see Sect. 6.2.1), which can substantially facilitate the company's work. Providing support for the company is hence one key pathway through which the industry level impacts the company level.

Financial resources are the final key element underlying the management of standards and regulation that we identify in our data. Here, we see a contrast between established companies and the smaller start-ups whose limited financial resources constrain their ability to participate in standardisation and lobby for changes in regulation.

6.1.1.2 Strategic Orientation and Organisational Support Structure

The three key characteristics of companies identified above determine to what degree they are able to orient their standards- and regulation-related work strategically. Our observations in Sect. 4.1.3 suggest that companies with little awareness, expertise, and financial resources tend to take a less strategic and more ad hoc approach. We therefore infer that these elements' presence is a necessary condition for a strong(er) strategic orientation. This manifests itself in aspects of the management, such as the degree to which standardisation activities are coordinated across the company and planned in advance.

This strategic orientation also forms the basis for an organisational support structure, which helps ensure that the innovation is systematically developed in line with requirements. An important function of this structure is assigning responsibilities both for operational management of standards and regulation, and for coordinating these activities across the company. In all interviewed companies, responsibility for operational tasks was tightly linked to the engineers developing a product. This appears to be good practice because of these tasks' technical nature and the close relationships between technical development work and standardisation/regulation efforts (see Sects. 4.2 and 6.1.2).

In companies with a strong strategic orientation, the organisational support structures also encompass clearly defined responsibilities for tasks related to planning and coordinating standardisation/regulation-related work.[1] In our case companies, these roles were attached to various organisational functions, including the new product development, regulatory affairs, and certification departments. Our data does not indicate that any of these affiliations is preferable per se, as long as the staff fulfilling this role are sufficiently influential within the company. Furthermore, companies can strengthen this organisational support by investing additional resources in full-time staff and tools supporting their work, such as the database tracking expertise related to specific standardisation/regulation topics that we observed at one company.

6.1.2   Activities for Managing Standards and Regulation

The factors discussed in Sect. 6.1.1 provide the basis for effectively managing standards and regulation in the innovation. The activities (depicted in the top half of Fig. 6.2) can be grouped into (1) core activities that are directly related to new product development (identifying regulation and standards, specifying the product, evaluating conformity to requirements) and (2) activities related to engaging at the industry level.

6.1.2.1   Core Activities: Identifying Regulation and Standards, Specifying the Product, Evaluating Conformity to Requirements

Based on the data outlined in Sects. 4.2.1, 4.2.2, and 4.2.3, we identify three core activities for managing standards and regulation which are part of the new product development process: (1) identifying applicable regulations and standards, (2) specifying the product, and (3) evaluating the product's conformity to the requirements. Carrying out all three in some form is necessary to ensure that the final product conforms to all applicable requirements. Nevertheless, we observe variation in how exactly firms pursue these tasks. This has implications for the degrees of freedom in new product development, as we outline below.

Before firms can take any action towards addressing standards and regulation in their NPD process, they need to know which requirements apply to their product, making *identifying regulation and standards* an

---

[1]Companies with an ad hoc approach tend to limit themselves to the operational tasks and therefore do often not address these duties in their support structures.

essential task. Our observations suggest that companies should do so at a very early stage, possibly already when deciding whether to invest in a new technology. This enables them to shape their product in a way which meets the requirements from the outset. Firms need to continue identifying requirements throughout the NPD process because rules are subject to change, and because not all technological aspects where standards/regulation apply may be foreseeable at the outset of the NPD process.

We also observe that not all companies are able to do so on their own, due to lacking awareness and expertise. This may result in an ad hoc approach to the topic and missing organisational support. However, such firms can rely on supporting institutions from the industry (see Sect. 6.2.1) to 'outsource' this activity and rely on third parties (e.g. consultants, notified bodies, and—in the case of component suppliers—clients) to identify relevant requirements on their behalf. However, our case shows that doing so has two drawbacks for the subsequent activities: (1) In some situations, companies may have discretion over which standards and regulation that they apply to their innovation, e.g. when multiple directives could be applied. To take advantage of this opportunity, they need to be aware of potential alternatives and evaluate the alternatives' consequences. (2) Relying on an external party to stay informed about changing requirements may delay the point in time when companies learn about new developments. Consequently, all companies in our case that followed a strategic approach to managing standards and regulation emphasised the importance of identifying regulations and standards for the subsequent activities.

The requirements identified in this first step are fed into the process of *specifying the product*, which includes 'translating' the contents of standards and regulation into concrete requirements, and designing the product in such a way that it meets these requirements. The case shows that especially requirements related to safety often take a very high level of expertise to implement and consequently all interviewed companies relied to some degree on external expertise in this step, and also used standardised components which were proven beforehand to meet the requirements. This activity therefore, again, benefits from a well-developed industry structure with supporting institutions (see Sect. 6.2.1).

Finally, companies need to *evaluate their product's conformity to the requirements* as part of the NPD process. Our case shows that firms should ideally carry out a first evaluation when deciding whether to invest in a technology and then repeat the assessment at regular intervals throughout the process. An initial appraisal of the innovation's potential to conform to the requirements enables companies to estimate the needed effort to address the topic in the NPD process and—in the worst case—prevents them from investing in technologies that cannot be marketed due to barriers discussed in Sect. 3.5. A firm's ability to effectively conduct such an initial appraisal relies on its strategic orientation, because of the understanding needed to assess factors, such as the likely impact of standards and regulation and their potential future developments.

Once companies invest in developing a technology for which standards and regulation are relevant, the case suggests that they should regularly review its conformity, potentially with the help of industry-level supporting institutions if the company's own expertise is insufficient. Doing so throughout the process reduces the need for duplicating development work if the results are fed back into the product specification process in a timely manner.

6.1.2.2 Engaging in Standardisation and Regulation

Engaging in standardisation and regulation is an additional, optional outward-looking activity (see Sect. 4.2.2), which provides the main path for companies to influence their environment. The examples of the smaller start-up manufacturers in our case show that developing a product which is acceptable for the market is possible without directly influencing standards and regulation. However, doing so opens up additional opportunities because it allows companies to contribute to developments on the industry- and wider context levels and provides them with the additional option of attempting to adapt standards and regulation rather than the innovation when conforming to them is impossible or difficult (see Sect. 3.5).

These activities rely heavily on a strong foundation (see Sect. 6.1.1) because they are relatively resource- and knowledge-intensive (both in terms of money and expertise), and also require the company to adopt a strategic outlook on the technology. The hurdles for mCHP's market introduction would most likely have been too high (locking the technology out of the market) if none of the companies had taken the initiative to develop standards and influence regulation. Although this is clearly a benefit of this engagement, actors who did not contribute also benefit to a large extent from the results (see Sect. 5.3). This implies that companies need a high degree of strategic vision and long-term thinking, aiming to develop a 'large pie for everyone' rather than a 'small pie for themselves' (at the risk of 'having no pie at all'), to invest in influencing standards and regulation for a new technology. Such long-term thinking, both within the company and at industry level, is also needed to successfully navigate the dynamic processes related to this topic (see Chapter 5, Sects. 6.2 and 6.3).

6.1.3   *Degrees of Freedom for New Product Development*

The aspects outlined so far have strong implications for the degrees of freedom for developing a new product. Depending on how they are handled, companies may enjoy a large scope for developing their own solutions or may be somewhat more restricted in key areas.

The company in our case that perceived standards mainly as limiting its freedom in developing mCHP (see Sect. 4.2.4) is also the one that was the least invested in the activities outlined above and relied to a very large degree on notified bodies and consultants (also see Table 4.1). Even though the interviewee at this company commended the notified body for its flexible approach in conformity assessment, the company's relatively low level of activity made it more dependant on external parties. This may have contributed to reducing the room to implement its own solutions.

The data clearly shows the benefits of taking an active approach towards the tasks outlined above. By doing so, firms can create a substantial amount of 'space' for innovating. In particular, three factors explain how this 'space' can be created: (1) The leeway in identifying regulation and standards (see the discussion earlier in this chapter and Sect. 4.2.1), (2) the open nature of many standards and different ways of demonstrating conformity (see Chapter 3), and (3) the potential to influence standards and regulation (see the discussion above and Sect. 4.2.2 and Chapter 5). Companies in the case who managed the topic strategically combined these factors in various ways (see e.g. the example of bringing new methods for ensuring product safety into the standard in Sect. 4.2.4) in order to develop innovative solutions while ensuring the final product's fit to the requirements. Consequently, all interviewed actors who followed such an approach agreed that they enjoyed a relatively large degree of freedom for developing the innovation while benefitting from the relatively stable basis offered by standards and regulation described in Chapter 3.

6.2   INDUSTRY LEVEL STRUCTURE AND PROCESSES

Following the theoretical analysis of the company-level management in the previous chapter, we now turn our attention to the industry level. Activities on the industry level are likely to focus on the standards which have the strongest impact on an innovation. In highly regulated markets, these standards are often linked to regulation (see Chapter 3).

Figure 6.3 summarises our findings regarding the work at the industry level. Again, we observe a number of underlying factors which contribute to an industry structure that facilitates activities in which standards and regulation are addressed (see bottom-half of Fig. 6.3 and Sect. 6.2.1). These activities are shown in the top of Fig. 6.3 and

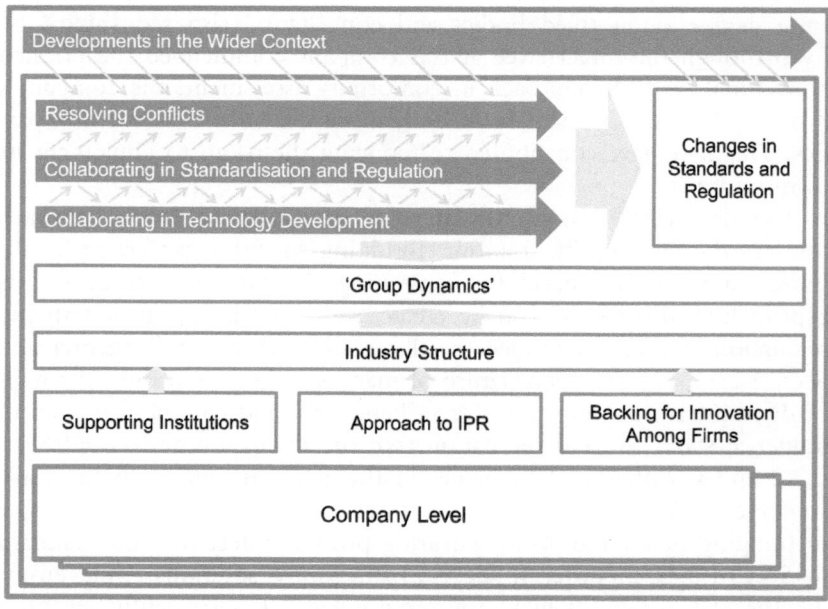

Fig. 6.3 Industry-level structure and processes for addressing standards and regulation

discussed in detail in Sect. 6.2.2. Furthermore, developments in the wider context influence the industry-level activities and vice versa, as we show in Sect. 6.2.2 and discuss in more detail in Sect. 6.3.

6.2.1   Key Elements of the Industry Structure

Our case clearly shows that the industry-level activities happen on the background of certain industry structures that may support (as we observed) or hinder the process. While the industry structure obviously consists of many elements, most of which are beyond the scope of this study, the data presented in Sect. 5.1 reveal three fundamental elements: supporting institutions, approach to IPR, backing for the innovation among firms (shown at the bottom of Fig. 6.3). These elements explain much of the success that we observe in our case. Below, we elucidate them and show how they contribute to an industry structure that is conducive to addressing standards and regulation for an innovation. We also briefly consider how such an industry structure can emerge.

6.2.1.1 Fundamental Elements: Supporting Institutions, Approach to IPR, Backing for Innovation

First, throughout our data in Chapters 4 and 5 it becomes apparent how crucial a number of *supporting institutions* were for all aspects of the case. Their influence extends to company-internal management (as discussed in Sect. 6.1), industry-level collaboration, and attempts to influence standards and regulation. Table 6.1 summarises the supporting institutions which we encountered in the mCHP case and the functions that they fulfilled.

The list of institutions and functions in Table 6.1 is specific to our case and therefore unlikely to be exhaustive. For example, it is conceivable that NGOs could support an innovation with social and/or environmental benefits, and contribute to the management of standards and regulation by influencing policy makers and the public debate in the wider context (see Sect. 6.3) in that technology's favour. Although the composition and functions of supporting institutions are case-specific, presence of such institutions in general is likely to be important in managing the co-evolution of innovation, standards and regulation. Our case suggests that these supporting institutions' contribution to the process is even larger than the sum of the individual functions listed in Table 6.1. One reason for this is these institutions' lack of a direct (financial)

Table 6.1 Overview over functions fulfilled by supporting institutions in the mCHP case

Supporting institution	Functions
External consultancy	• Provide technical expertise and knowledge about applicable regulation and standards • Represent individual companies or the entire industry in standardisation committees
Notified bodies	• Provide knowledge about applicable regulation and standards • Enable market access by issuing conformity certificates • When harmonised standards are absent: translate 'essential requirements' into concrete criteria
Industry associations	• Provide a forum for industry actors to agree on common positions and 'talk with one voice' • Provide access to regulatory processes for the industry • Observe developments in adjacent areas of regulation and standardisation
Academic research institute	• Support field trials and other collaborations for technology development • Provide independent technical expertise in standardisation committees

interest in the technology's success, which lends the industry's claims and actions credibility. In addition to them facilitating much of the necessary work, both on the company- and industry level, they can therefore be seen as amplifying the impact of the innovators' own activities.

Second, we identify the *approach to IPR* as core to an industry structure which supports managing standards and regulation effectively. As we show in Sect. 5.1.4, actors in the case placed a high importance on IPR in technology development partnerships. However, they consciously decided to leave the topic out of activities directly related to standards and regulation. While the best way of handling IPR issues may be case-specific, our data shows that an industry needs to ensure that the chosen approach does not discourage others from joining the industry's efforts. Because collaborating in technology development and standardisation/regulation is key to the industry activities (see Sect. 6.2.2), the IPR regime must support them. This means that on the one hand all contributors' IP must be protected. On the other hand, no party should be able to use its IP for dominating the cooperation in a way that causes potential developers to refrain from or stop contributing to the technology. In addition, such domination by one party would likely also make the resulting standards unacceptable to other key stakeholders on whose support the innovation depends. Especially when these standards are linked to regulation (see Table 3.3), the approach to IPR must also be acceptable to regulators and other stakeholders. For example, standards which are used to specify essential requirements under the 'New Approach' should not incorporate IP that is subject to licensing. When addressing standards with no link to regulation, approaches to IPR that involve standard-essential patents (as commonly discussed in the literature, see Sects. 1.1 and 7.3.2) may be more acceptable.

The case suggests *backing for the innovation among firms* to be the third key element of the industry structure that determines to what extent the processes for addressing standards and regulation can be effective. Whether the majority key firms in the industry or only a few players support the innovation influences the extent of industry-internal conflicts, and how the innovation's legitimacy is perceived by outside actors. Furthermore, the degree of backing has ramifications for the 'group dynamics' that we discuss in Sect. 6.2.3.

6.2.1.2 Emergence of the Industry Structure

The three fundamental elements discussed above make up the parts of the *industry structure* that are relevant for the processes that we discuss in Sect. 6.2.2. When, as we observed in our case, these attributes are well aligned (i.e. a good network of supporting institutions is available, a fitting approach to IPR is employed, and there is widespread backing among firms) this structure provides a solid foundation for these processes. On the other hand, if some of the elements identified above are missing, this is likely to hinder the industry-level work needed to ensure alignment between the innovation and standards/regulation. In addition, such missing elements may have negative implications for company-level work.

Although our data does not offer detailed insights into how this industry structure has been built over time, it clearly is the result of a long-term development on which the companies were able to draw in the present case. Ultimately, this long-term development is likely to have been driven to a large extent by the individual companies in the industry who have been contributing to setting up supporting institutions, such as industry associations, and establishing an effective approach to IPR. Also the backing for the technology requires a long-term commitment, as our case shows. Individual companies can try enlisting their competitors in contributing to establishing these key fundamental elements, but are unlikely to succeed in building them on their own. Furthermore, some elements that can be leveraged in this context (e.g. NGOs as supporting institutions) may also appear without industry-actors' direct involvement.

6.2.2   Industry-Level Processes for Facilitating the Innovation

The elements of the industry structure outlined in Sect. 6.2.1 underlie the joint industry-level activities that eventually lead to changes in standards and regulation needed to support an innovation. In our case, we categorise industry-level activities (see Chapter 5) into three core processes: (1) *collaborating in technology development*, (2) *collaborating in standardisation and regulation*, and (3) *resolving conflicts*. As the case and our further discussion below show, it is essential for achieving the needed changes in standards and regulation that these processes are jointly driven by companies from the industry (unless one innovator is strong enough to 'push them through' alone), and that need to be coordinated well in order to deliver the desired results.

The findings from Sect. 5.1 suggest that *collaborating in technology development* both helps actors in the industry to jointly overcome technological challenges in some areas and also provides a basis for the further activities. Through their joint engagement in developing an innovation, actors in an industry (1) share a strong interest in the technology's success, (2) develop a common outlook on standardisation and regulation issues, and (3) can more easily address technological issues, that arise in the process of developing standards/regulation, together. These points also contribute to a tight link between technology development and *collaborating in standardisation/regulation*. For example, evidence created in technology development cooperation projects was directly used in discussions on standards with other stakeholders in the mCHP development process (see Sect. 5.2.1).

Both types of collaboration benefit from a well-developed industry structure (see Sect. 6.2.1). Supporting institutions facilitate the cooperation because they provide already established forums where the work can take place, help coordinate the activities, and provide expertise and access to policy makers. An appropriate approach to IPR ensures that participating in cooperation is viable in terms of protecting one's own input while avoiding that certain actors can dominate the technology's development through their patents. Nevertheless, even when these factors are present, some conflicts may occur. Conflicts are particularly likely if important actors in the industry do not back the innovation (as could be observed in our case, see Sect. 5.1.3). Furthermore, the developments in the wider context about which we theorise in Sect. 6.3 may also contribute to conflicts, as could be observed in our case. This makes *resolving conflicts* a final key activity on the industry level to ensure that the changes in standards and regulation needed for an innovation can be achieved. Also for this key activity, our data shows the industry structure's importance for this issue, with supporting institutions playing key roles in helping to solve these issues (see Table 6.1).

6.2.2.1 Individual Companies' Contribution to Industry-Level Processes

The industry-level processes are chiefly driven by individual companies' contributions. Although the case shows that these processes often last several years and companies need a strategic long-term view to navigate them effectively, their results are much more immediate than building the industry structure outlined earlier. Furthermore, the industry-level processes enable companies to collaborate on those activities that are needed to align the technology, standards, and regulation, which cannot be carried out at company-level. Especially for companies which have insufficient clout on their own for driving changes in standards/regulation and engaging with the wider context (see Sect. 6.3), contributing to these processes is the key path to influencing developments at the industry- and wider context levels.

6.2.3 'Group Dynamics' in the Industry

As we observed in Sect. 5.1.3, the industry structure and collaboration processes in the mCHP resulted in certain 'group dynamics'. In our case, the strong support among industry and the obstacles to implementing the innovation, which were perceived in common across most involved actors, created mCHP's backers forming a very closely-knit group. They adopted a strong 'us vs. them' mentality when dealing with any parties not supporting the innovation. On the other hand, a lack of support and conflicting perceptions of the technology's environment may result in very contentious 'group dynamics'.

Our case shows that such 'group dynamics' cause the involved companies to adapt a common outlook on the technology and what was needed to make it successful. Consequently, in such a setting, few disagreements between firms are likely to occur and the processes for resolving conflicts are mainly needed in dealing with the wider context instead of addressing industry-level issues. This common outlook and 'us vs. them' mentality also enables an industry to speak with one voice when addressing topics in the wider context.

However, on the other hand such a closely-knit group of actors also may have drawbacks. First, it may endanger the industry of entering a 'groupthink' mode of acting. More importantly, it may impact on how the industry is seen by stakeholders in the wider context. 'Group dynamics', such as the ones observed in the mCHP case, carry the risk that the industry is perceived as a colluding group, which writes its own rules and engages in regulatory capture. Our data does not show whether mCHP's backers were indeed perceived in this manner, but the discussion on how to interpret the industry's own energy efficiency calculation method in the wake of the Volkswagen Diesel scandal (see Sect. 5.2.2) shows that some actors were aware of this risk. Potentially, the credibility given to the technology by some of the supporting institutions (see Sect. 6.2.1) may also counter-act this threat, although more research is needed to investigate this.

Despite these possible pitfalls of acting as a too closely-knit group on the industry level, our case suggests that doing so generally supports the industry-level processes. The benefits of reduced conflicts and 'speaking with one voice' are potentially substantial and supported mCHP's development considerably. The collaborations to develop the technology and in particular the successful handling of the European Commission's intervention in the energy-labelling issue would have been hampered by other possible constellations of actors. Similar benefits are also likely to apply to other cases.

6.3   Developments and Associated Processes in the Wider Context

As a final area within the three levels of our framework (see Fig. 6.1), our case shows the importance of *developments in the innovation's wider context* beyond the industry, and the associated processes of managing them.

All our interviewees repeatedly stressed the importance of managing links with interests and actors outside the industry, such as regulators and developers of other technologies. Furthermore, our data reveals the aspects of standardisation related to the wider context to be both the most contentious topics in the mCHP case, and the ones demanding the most attention of the innovators (see the introduction to Sect. 5.2).

In the mCHP case, we observed three such important developments, which also were intertwined at some points: (1) One related to changes in access to the electricity grid, (2) trajectories of other innovations that were emerging simultaneously in that space (e.g. renewable energy generation, see Sect. 5.2.1), and (3) events related to political agendas and policy objectives that drove regulators' activities (e.g. reducing CO_2 emissions and promoting renewable energy, see Sect. 5.2.2). In addition, several interviewees expected trends relating to re-use, recyclability and reparability (RRR) to become similarly impactful in the future. Beyond these examples, other types of developments could play similar roles in other cases. For example, both important societal debates,[2] and scientific findings on risks associated with an innovation[3] could have substantial implications for a technology's standards and regulation. Overall, these types of trajectories in the wider context are therefore highly relevant elements for theorising as part of the three levels in our framework.

Our case offers a clear picture of how these developments interact with the activities on which we focus in this study. While the case does not provide detailed insights into these trajectories themselves, it does thus offer an excellent basis for theorising about their interactions with standards in an innovation's development. Figure 6.4 shows these interactions and provides a more detailed look at the link between the industry level and the wider context shown in the topmost part of Fig. 6.3.

In Sect. 6.3.1, we discuss the relevance of these developments further and shed light on their effects on an innovation's development. We then theorise in Sect. 6.3.2 about strategies that actors in an industry can use to influence developments in the wider context.

---

[2]The societal debate following the revelations regarding the automotive industry's emission-testing practices can be seen as an example of this.

[3]For example, scientific findings about certain medical treatments' effectiveness may have implications for standards and regulation concerning innovations in drugs for these treatments.

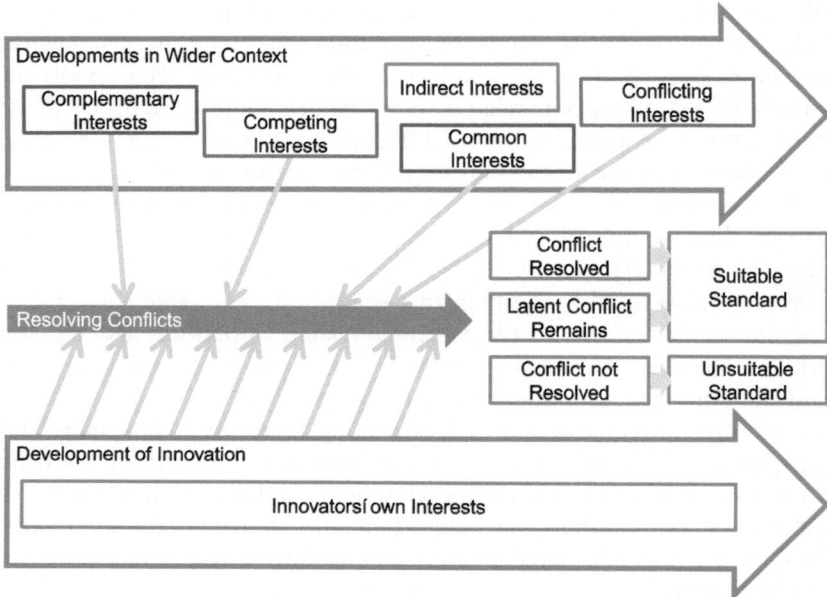

Fig. 6.4 Interactions between the innovation and developments in the wider context

6.3.1   Relevance and Effects of Developments in the Wider Context

The types of trajectories outlined above are driven by interests, which in many cases may not be aligned with the needs of a specific innovation, and can directly lead to new requirements. For example, the data presented in Sect. 5.2.1 shows how designers of renewable energy generation technologies and grid operators drove changes to grid connection standards with which mCHP had to comply. In terms of standards that innovators may encounter (see Sect. 3.5), such processes in the wider context are by definition always relevant for standards that relate to regulation (which is made by policy makers and other actors who are part of the wider context). However, work in areas with no link to regulation may equally be impacted by the wider context, for example when standards define interfaces to a larger system, such as the electricity grid in the mCHP case.

Table 6.2 Examples of different types of interest in interactions with the developments related to electricity grid access in the mCHP case

Type of interest	Example of Interest	Actor(s) holding the interest
Innovators' own interest	Secure access to electricity grid for mCHP appliances	Developers of mCHP
Common interest	Gain access to electricity grid for small generators	Producers of renewable power generators
Complementary interest	Shift balance of electricity generation away from large competitors	Small electricity providers
Competing interest	Allow wider frequency bands in electricity grid	Producers of renewable power generators
Conflicting interest	Retain easily manageable grid by keeping small generators out	Grid operators
Indirect interest	Exit nuclear power	German government

Such external influences can be positive or negative for the innovation, and may therefore ultimately lead to conflicts. This depends on the interests that are at stake. In our case, we identify six relevant types of interest (see Table 6.2 for one example of each from the interactions concerning mCHP's grid-access[4]). (1) Innovators have their *own interests* in how the wider context should develop. (2) These interests may be shared with other actors who have a *common interest*. (3) Actors may also have *complementary interests*, which can be supported by developments that are in line with the innovators' own interest. On the other hand, there may be (4) *competing interests* which aim to achieve an outcome that is incompatible with the innovators' needs. Finally, there may be (5) *conflicting interests* that collide head-on with the innovators' goals. In addition, there may be (6) *indirect interests*, which are only indirectly linked to achieving outcomes in the wider context that support the innovation.

As the examples in Table 6.2 show, the interests and associated actors that are involved in the industry's wider context are likely to be highly diverse, making the developments that take place there very dynamic. Depending on how these interests are distributed among the actors in the wider context, these developments may be contentious issues. This

[4]Details can be found in Sect. 5.2.1

requires an innovation's supporters to adopt a careful approach, as we outline in the following chapter.

6.3.2   Influencing Developments in the Wider Context

The kinds of development outlined in above are often embedded in major movements, such as the efforts to reduce CO_2 emissions. They may involve many stakeholders with diverse interests from different industries, governments, NGOs, consumers, and other actors. Also the logics of change in different wider context s vary and may not always be completely transparent, as the interaction with the European Commission in our case shows (see Sect. 5.2.2).

Consequently, innovators tend to hold relatively little sway over external developments, although the exact extent to which they can influence them is case-specific. For example, the developers of mCHP had a much smaller influence in developing standards for access to the electricity grid than when handling the requirements for energy labelling (see the data in Sect. 5.2). Within the bounds of this influence, innovators can take an active approach to managing these developments as part of the process of *resolving conflicts* (see Figs. 6.3 and 6.4). Our case exhibits four basic strategies that can be used as part of such an active approach, which we summarise in Table 6.3.[5]

These four strategies are not mutually exclusive. They can be used in parallel, even for influencing one development in the wider context, as the interactions with the developments regarding grid-access standards in our case show. This reflects the multitude of interests and associated actors involved that we outlined in Sect. 6.3.1. Each of the four strategies has certain prerequisites, which to a large extent relate to interests of other actors and the structure of the wider context (see Table 6.3). Actors with common or complementary interests can therefore be involved in coalitions, whereas competing and conflicting interests may be addressed by lobbying (if the associated actors are open to discussions) and/or adapting the technology accordingly. Furthermore, actors with competing and conflicting interests may sometimes also not be able to act on these interests. In these cases, persisting with own preferences may be an appropriate course of action.

[5] Again, this list may not be complete and other potential strategies, which we did not observe in our case, may exist.

Table 6.3 Strategies for influencing developments in the wider context

Strategy	(1) Form coalitions with actors in wider context	(2) Lobby actors in wider context	(3) Adapt innovation to developments	(4) Persist with own preferences
Prerequisites	Actors with similar interests exist in wider context	Convincing evidence about the technology is available Actors in wider context open to discussions	Value proposition not compromised significantly by meeting requirements	Structure of wider context presents opportunity to follow own approach
Consequences	Partners support shaping developments in the innovation's favour	Actors in wider context may be more receptive to the innovation	May lead to goodwill from actors in wider context	Potential additional conflicts with actors in wider context
Examples in mCHP case	Partnerships with small electricity suppliers (Sect. 5.2.1)	Convincing grid operators about the innovation's special characteristics (Sect. 5.2.1)	Adapting technology to cover wider grid frequency bands (Sect. 5.2.1)	Efficiency calculation method (Sect. 5.2.2)

Through the consequences named in Table 6.3, the four strategies contribute to the outcome of innovators' attempts to resolve conflicts. Three such results are possible: (1) In the best case, conflicts with actors in the wider context are resolved, leading to the development of standards that are suitable for the innovation (i.e. standards with which the innovation can conform, see Sect. 3.5). In our case, we observed this outcome in many technical areas which were key for grid-access where small generators could eventually be connected to the electricity grid (see Sect. 5.2.1). (2) In addition, suitable standards can be developed following innovators persisting with their preferences. In this situation, which we observed in our case on the efficiency calculation issue (see Sect. 5.2.2), latent conflicts with other actors in the wider context may remain. Even though this outcome initially supports the innovation's market introduction, any latent conflicts may re-emerge later on and potentially lead to new problems. For example, in resolving the questions related to the calculation method in our case it was initially unclear how market surveillance authorities would treat the industry's use of its own standards instead of the European Commission's method and whether this would lead to further issues. (3) Finally, industry actors may also fail to resolve conflicts to their satisfaction and face resulting standards with which the innovation cannot easily conform. As we observe on the issue of grid frequency (see Sect. 5.2.1), this is a likely outcome for issues where there are insufficient actors in the wider context with whom alliances can be formed and competing/conflicting interests are too strong.

In conclusion, developments in the innovation's wider context are driven by a large variety of actors with diverse interests that may favour an innovation or oppose it. Depending on how these interests are eventually balanced, this context can boost an innovation or pose substantial barriers. Innovators tend to have limited influence on the wider context, which also depends on factors like the interests at stake, and the logic according to which changes in a development happen. While avenues for actively influencing these developments are available, their success ultimately depends on the characteristics of the specific development.

6.4   Final Thoughts on Our Grounded Theory

In the introduction to this chapter and Fig. 6.1, we claimed that innovators' activities on the company-, industry-, and wider-context levels need to be concerted in order to achieve alignment between an innovation and the applicable standards/regulation. Our discussion shows this to be true. While an innovation is ultimately driven by individual companies that develop the technology, any needed changes in standards and regulation require action on the other levels. We already expected the link between the company- and industry levels but also discovered the significance of the wider context.

As our theory shows, these links mean that the processes which we study are not linear but highly dynamic. They depend on the input of a large variety of actors, in addition to the companies developing the innovation. These actors may have very different stakes in the innovation and diverse functions to fulfil. These functions include, for example, industry associations providing forums for collaboration and supporting lobbying efforts, governments offering stability for the innovation, or consultants and researchers supplying expertise in key areas. Furthermore, not all actors involved in the process may be in favour of the innovation. This poses some of the most significant challenges for aligning the innovation, standards, and regulation.

Beyond this, our findings also mean that aligning the innovation with standards and regulation is not a goal in itself. The mCHP case shows that doing so may often be a necessary condition for introducing a technology into the market. Additionally, the observations in Sect. 6.3 suggest that the function of standards and regulation goes much further. Arguably, standards and regulation fulfil a key function of translating the large trends and needs in a technology's wider context (e.g. reducing CO_2 emissions, building a stable electricity grid) into concrete technical requirements for a product. This means that aligning an innovation with standards equally contributes to aligning the innovation with the demands of key actors in the wider context on whom it ultimately depends for its success. The theory, which we have built based on the evidence from the mCHP case, offers guidance on how this can be achieved. This makes our theory a theory at the core of developing an innovation, going beyond the theory about managing standards that we anticipated building when we initiated this study.

CHAPTER 7

Conclusions: Managing Innovation and Standards Within the Company and Beyond

Abstract This chapter concludes the book by discussing the findings in light of literature and giving clear managerial advice. The study extends the literature on effects of standards and regulation on innovation, integrating them into new product development, and associated dynamics on the industry level. In addition, the results link to other streams of literature, such as theories about sociotechnical systems, regulatory uncertainty, co-opetition, and the need for rules in the functioning of markets. This chapter shows these links and outlines trajectories for future research to explore them further. It also highlights the study's managerial implications and translates them into clear advice for innovators and other actors, such as industry associations.

Keywords Effects of standards on innovation · Managing standards in innovation contexts · Standardisation · Regulation Innovation management · Industry dynamics

In this study, we aimed to develop a grounded theory about innovative companies' management of the critical implications that standards and technical regulation have for developing new technologies. In Chapter 6, we detail the core concepts (three levels at which various activities occur), which make up this theory, and the relationships between them. This concluding chapter highlights the theory's contribution to literature

© The Author(s) 2019   139
P. M. Wiegmann, *Managing Innovation and Standards*,
https://doi.org/10.1007/978-3-030-01532-9_7

(Sects. 7.1–7.3), main managerial implications (Sect. 7.4), and implications for future research (Sect. 7.5).

A first contribution of our study therefore lies in the new insights it provides into the effects of standards on innovation (see the discussion in Sect. 7.1). It clearly demonstrates their critical implications and provides new insights into some of the causal mechanisms behind the effects. In order to address them, our study shows that managers need to align the innovation with the relevant standards by adapting the technology, standards, and/or regulation. Our grounded theory approach revealed that this 'managing', which motivated our interest in the topic, does not only happen on the company level. In addition, processes that happen beyond the company at the industry level and in the wider context turned out to be more important than expected. We can therefore relate these findings to Van de Ven's (2005) concepts of 'running in packs' and 'political savvy'. Furthermore, while our study focuses on the 'managing', it also links to related topics like sociotechnical systems (e.g. Geels, 2004; Smith & Raven, 2012; Smith, Voß, & Grin, 2010), and the functions of standards and regulation in establishing markets (Polanyi, 2001).

At the outset of our study, we identified three important gaps in the existing literature (see Sect. 1.2.4) addressing our research question about managing standards, which guide our subsequent discussion: (1) a lack of attention to activities at the firm level, (2) few findings about companies' interactions with the industry level, and (3) limited findings about industry-level dynamics. Our study's detailed findings and open insights allow us to contribute to closing all three gaps. In addition, our study also highlights the importance of dynamics that are associated with the innovation's wider context. In Sect. 7.2, we discuss our theoretical contribution on the company level. Sect. 7.3 addresses the dynamics that affect the industry level and wider context.

7.1   STANDARDS' EFFECTS ON INNOVATION

As we show throughout our study, standards have very profound effects on innovation. Our contribution to the literature on these effects is threefold. First, we show the causal mechanisms behind these effects and demonstrate the importance of coherent sets of standards for an innovation (Sect. 7.1.1). Second, we add to existing findings on the circumstances under which standards are likely to have the strongest effects on innovations (Sect. 7.1.2). Finally, we identify the lack of standards as a key source of ambiguity and uncertainty for an innovation (Sect. 7.1.3).

7.1.1 Existing Standards' Effects on Innovation

In Table 1.1, we summarised extant findings on how standards can support and/or hinder innovation. Our study adds to these findings by providing more detailed insights into causal mechanisms behind the effects already identified by the current literature. In particular, legitimacy and market access (see, e.g. Borraz, 2007; Botzem & Dobusch, 2012; Delemarle, 2017; Tamm Hallström & Boström, 2010) and creating supporting infrastructures (see Teece, 1986, 2006) are key to our study and illustrated in much detail by our case. Furthermore, the mCHP case exemplifies other effects found in extant literature, e.g. standards being an important information source for NPD activities (see, e.g. Allen & Sriram, 2000; Blind & Gauch, 2009; Egyedi & Ortt, 2017; Featherston, Ho, Brévignon-Dodin, & O'Sullivan, 2016; Van de Ven, 1993) or their role in specifying testing and performance requirements (see Abraham & Reed, 2002; de Vries & Verhagen, 2016; Swann, 2010).

Interestingly, some of the effects outlined in Table 1.1 and Sect. 1.1 were not recognised by the experts in our interviews. For example, literature (e.g. Kondo, 2000; Tassey, 2000) states that standards limit available options for innovation. Most interviewees clearly stated that standards as such did not prevent them from any choices that they deemed beneficial for the technology and left considerable degrees of freedom for innovating (see Sects. 4.2.4 and 5.3). What they did criticise was particular standards posing difficult requirements or reflecting strategic moves by other actors who were attempting to use standards for blocking the technology (also see Sect. 7.3). This shows that at least some of the effects identified in the literature (both positive and negative) do not apply to all standards per se. Instead, whether a particular standard has positive or negative implications for an innovation depends on that standard's contents. In particular, it depends on whether the innovation can be designed in such a way that it conforms to the standard (see Sect. 3.5) and how easily this can be done.

While each distinct standard that touches on an innovation is relevant on its own in this context, our study and existing literature (Featherston et al., 2016; Ho & O'Sullivan, 2017) show that innovations can depend on large sets of standards. Innovations therefore do not only depend on a small number of individual standards but often must incorporate requirements laid down in a variety of standards. Even for a relatively simple technology like mCHP (compared to systemic innovations like autonomous driving or Smart Cities), this set encompasses a substantial number of standards coming from all categories in Table 3.3 and covering multiple economic functions (see Blind, 2004, 2017; Egyedi & Ortt, 2017; Swann, 2010). Even more extensive arrays of standards are likely to become relevant for technologies that are more complex. In many cases, these sets may include different standards formulating requirements for related aspects of a product and/or standards that relate and build on each other. This underlines the need for coherence among standards (see de Vries, 1999; Featherston et al., 2016; Ho & O'Sullivan, 2017) and architectures on which individual standards are based (see, e.g. van Schewick, 2010) in order to realise their potential positive effects.

Overall, our study suggests that the positive effects of standards on innovation by far outweigh the negative ones. The case clearly shows that standards not only impact on innovation positively in many ways, but may even be a necessary condition for bringing a new technology to the market. This also relates to our observation in Sect. 6.4 that standards fulfil the important function of specifying technological requirements that result from needs of actors in the wider context.

There is some previous standardisation literature which relates to this observation: Delemarle (2017), Botzem and Dobusch (2012), and Van de Ven (1993) discuss the role of standards in forming markets and legitimising innovations. Tassey (2000, p. 588) describes standards as "a balance between the requirements of users, the technological possibilities (...) and constraints imposed by government for the benefit of society in general". De Vries and Verhagen's (2016) case of energy performance standards for houses shows how standards that impact on innovation can directly result from demands associated with trends in a technology's wider context. Nevertheless, despite Geels's (2004) recognition of the function that standards fulfil in technological transitions, extant standardisation literature does not explicitly link to this literature. Our observations suggest that standards may fulfil a role in facilitating technology transitions by helping to define technological niches and providing protective space (see Smith & Raven, 2012; Smith et al., 2010).

7.1.2   Strength of Standards' Effects on Innovation

While all standards that are relevant for an innovation have some impact, our study also shows that the strength of this impact differs across standards. Several such factors can already be derived from the existing literature: Multiple authors (e.g. Blind & Gauch, 2009; Tassey, 2000) argue that the progress of the technological trajectory at the point in time when a standard is developed influences the standard's eventual effect on the innovation. Tassey (2000) also points out that 'design-based' standards have potentially much more profound constraining effects than 'performance-based' standards (see Sect. 1.1). Another factor mentioned in this context is the degree to which a technology is subject to network effects and switching costs which determines the degree to which lock-in poses issues for innovations (e.g. David, 1985). Based on the types of standards that we encountered (see Table 3.3), we add the strength of the link between a standard and regulation as a factor that amplifies both potential positive and negative effects of the standard.

Increases in positive effects driven by standards that support regulation mainly relate to an innovation's market access. In this context, support from standards goes beyond legitimising innovations in the eyes of potential users and other stakeholders (as already discussed by, e.g. Botzem & Dobusch, 2012; Delemarle, 2017; Tamm Hallström & Boström, 2010). Our study shows that close connections between standards and regulation facilitate the proof of an innovation's regulatory compliance substantially and provide additional (legal) certainty to innovators and other stakeholders alike. Such standards therefore arguably enable the innovation being offered in the market in the first place.

On the other hand, closer links between a standard and regulation also make implementing solutions that do not conform to the standard more difficult (e.g. because of expensive documentation and testing procedures to prove such solutions' equivalent performance). Particular standards which might hinder an innovation therefore become difficult to avoid or de facto compulsory in this situation. Whereas a hindering standard with no link to regulation only requires an innovator to invest in developing an alternative solution and/or find other ways of legitimising the product, a hindering standard with strong links to innovation may effectively lock a product out of the market.

7.1.3   Uncertainty Resulting from Missing Standards

All of the above assumes that the contents of standards are known. However, our study shows that this is not always the case and relevant standards may not yet exist at a point in time when they are needed to support the innovation. As far as we are aware, in the current literature only Blind and Gauch (2009) offer insights about the effects of standards being unavailable when needed for a technology's further development. In particular, they find that missing terminology standards contribute to a proliferation of heterogeneous terminology. Our study goes further by clearly showing that lacking standards are a core source of uncertainty for both innovators and other stakeholders (users of the innovation, component suppliers, complementors, etc.), similar to the ambivalence resulting from regulatory uncertainty (see Hoffmann, Trautmann, & Schneider, 2008). This therefore underlines the argument that markets need clear rules guiding actors within them (Fligstein & McAdam, 2012; Polanyi, 2001).

Such unavailable standards lead to a multitude of ambiguities for innovation, such as unclear requirements for the technology, risks of supporting infrastructures not fitting the product, and users not understanding its benefits. These ambiguities are further amplified by the importance of the entire set of standards that applies to an innovation (see Sect. 7.1.1). For any missing standard in such a set, aspects like how it will relate to other standards once it emerges, which economic functions it will fulfil, or where it will fall into our taxonomy may be unknown a priori. Such missing standards therefore impact on all stages of the innovation's development, including conceptualising the product, working with suppliers and others on the technology, and introducing it in the market. Once all relevant standards are known, much of this ambiguity is resolved. Although standards are subject to change under some conditions—as both this study and previous literature (Egyedi & Heijnen, 2008; Wiegmann, de Vries, & Blind, 2017) show—they resolve this instability and uncertainty that would otherwise hinder innovation.

7.2   Managing Standards, Regulation, and Innovation

Extant literature extensively documents the substantial effects of standards on innovation (see Sect. 1.1), yet it offers few insights about how companies can manage this important topic. Extant literature on company-internal standardisation management mainly addresses companies' engagement in standardisation (e.g. Axelrod, Mitchell, Thomas, Bennett, & Bruderer, 1995; Blind & Mangelsdorf, 2016; Jakobs, 2017; Wakke, Blind, & De Vries, 2015), and the implementation of standards within companies (e.g. Adolphi, 1997; Foukaki, 2017; van Wessel, 2010). However, Großmann, Filipović, and Lazina (2016) are—to our knowledge—the only researchers who address managing standards in the context of innovation. Furthermore, the literature on standards mostly omits the link to regulation that we show to be essential in many situations. Our grounded theory model of managing standards and regulation at the company level (see Fig. 6.2 and Sect. 6.1) contributes findings that add to the literature on both counts.

Some aspects of these findings resemble existing theory about managing standards, showing that it also extends to the specific context of innovation. For example, our model distinguishes between short- to medium-term activities needed to address standards and regulation, and a number of supporting factors that enable these activities. This resembles the distinction between long-term governance and short-term management activities in van Wessel's (2010) framework, although the elements that make up these categories differ.

On other aspects, our model significantly extends the extant theory on company-level management of standards, as we outline below. In particular, our discussion of our model's firm-level parts revolves around three aspects: (1) the company-level support structure for managing standards and regulation (Sect. 7.2.1), (2) firms' approaches to integrating standards and regulation into their NPD processes and these approaches' effects on an innovation (Sect. 7.2.2), and (3) their involvement in external developments through engaging in standardisation and related activities (Sect. 7.2.3).

7.2.1   Organisational Support for Managing Standards and Regulation

Existing literature already addresses some elements of the organisational support structure needed. Adolphi (1997) focuses to a large extent on how firms integrate standardisation into their functional divisions. Van Wessel (2010) highlights the need for governance, which includes elements such as investment decisions and defining strategies, to support day-to-day activities related to standards. Foukaki (2017) identifies distinct 'standardisation management approaches' in companies that drive much of the subsequent activities. In line with this, several authors (Adolphi, 1997; Foukaki, 2017; Großmann et al., 2016; van Wessel, 2010) highlight the need for a strategic approach to standardisation. Our study confirms this need. In our theorising (see Sect. 6.1.1), we clearly argue that a strategic orientation towards standards enables companies to build an organisational support structure that contributes to handling standards and regulation in NPD. Our results suggest that such a strategic approach allows companies to coordinate their standardisation activities across their business and exploit the long-term effects of standards. Beyond this confirmation of the need for a strategic orientation, our study makes two further contributions on organisational support for managing standards and regulation to the literature.

First, we identify awareness, expertise, and financial resources as necessary conditions for developing a strategic orientation towards standards and regulation. These factors are in line with the findings of de Vries, Blind, Mangelsdorf, Verheul, and van der Zwan (2009) and Foukaki (2017)[1] but we add further insights into *how* they contribute to successfully addressing standards and regulation. According to our findings, awareness of the topic's importance and expertise (in particular strategic) help companies to assess standardisation in light of their business model and innovation activities. These factors therefore help them formulate a standardisation strategy (also see Adolphi, 1997; Jakobs, 2017), which covers aspects such as engaging in external standardisation and lobbying, and identifying areas where existing standards can be used. In addition, financial resources are essential for deriving such a strategy because of the associated costs (e.g. for qualified staff and travelling), which often are beyond the means of smaller companies.

Second, we show how a strategic approach helps to build the organisational support structure that underlies day-to-day activities, which may sometimes even be underdeveloped in large, otherwise professionally run companies (see Großmann et al., 2016). In this context, Adolphi (1997) focuses on different models regarding where firms incorporate standardisation work into their functional structures. Our study suggests that

---

[1] Foukaki's (2017) study was not yet available when we conducted the literature review underlying our work. Interestingly, her cases also lead her to identify awareness as a core concept in standardisation management that has not been addressed in the mainstream academic literature.

the specific organisational function (e.g. the R&D or production department) to which these tasks are attached is of secondary importance. While we observe different approaches across companies in that regard, none of them appears to be preferable per se. Instead, clearly defined responsibilities for planning standardisation work and ensuring that the responsible staff have sufficient influence and authority to ensure that these plans are implemented appear to be important for providing optimal support.

7.2.2 *Integrating Standards and Regulation into the Innovation Process*

The organisational support discussed above enables activities related to integrating standards and regulation into the innovation process. On a very fundamental level, we distinguish between active and passive approaches. They somewhat resemble Foukaki's (2017) assertive and vigilant approaches to participating in standardisation, but go further because they also touch on aspects like product design and involvement of third-party consultants. Whether a company adopts an active or passive approach is likely to be driven by the commonly held image of standards and regulation within the firm (i.e. whether they are seen as a welcome support or a necessary evil). Companies which appreciate the value of standards are more likely to adopt a (pro)active approach. Such approaches can be implemented, e.g. in terms of using the available leeway regarding which standards and regulation to apply, or exploiting the open nature of many standards (see the data in Sect. 4.2 and our theory in Sect. 6.1.2 for details). Our results suggest that doing so can lead to substantial degrees of freedom for developing an innovation. We therefore question to some extent the commonly held view that "firms need to strike a balance between both flexibility and standardization" (Lorenz, Raven, & Blind, 2017, p. 29).

Instead, it appears to be a question of managing standards in such a way that they enhance flexibility rather than constrain it. As we explained in Sect. 1.2.1, existing literature on how this can be done is extremely scarce. We are aware of only one earlier study (Großmann et al., 2016) that explicitly addresses the management of standards during an NPD process. This study therefore forms a 'benchmark' against which we compare our findings.

Großmann et al. (2016, p. 322) (integrate standardisation-related activities into a model of a generic stage-gate NPD process (covering six stages from idea to market introduction), which shares the core activities needed with our model (see Fig. 6.2) but differs on how these activities relate to each other. They suggest two specific standardisation-related tasks that take place in parallel to the core sequence of innovation development activities: (1) 'screening standards', which takes place in parallel to the early phases of the product's development, and (2) 'participating in standard setting committees', which happens next to later stages. Both closely resemble activities that we identify in our model: 'identifying regulation and standards', and 'engaging in standardisation and regulation' (see Fig. 6.2). In addition, our model entails 'specifying the product' and 'evaluating conformity to requirements' as distinct necessary activities in this context. Großmann et al.'s (2016) model includes these activities within the regular stages of the core NPD process ('development', followed by 'testing & validation').

While we find similar necessary activities, our findings challenge the sequential approach of Großmann et al.'s (2016) model. Our theorising (see Sect. 6.1.2) shows that this is unlikely to work in situations which are characterised by factors such as uncertainty about future standards (see Sect. 7.1.3), technological learning by the company,[2] and attempts by actors in the technology's wider context to influence standards and regulation (see Sect. 7.3). These circumstances imply, among other things, that some relevant standards and regulation are not known at the outset of the NPD process and are continuously subject to change (see, e.g. Wiegmann et al., 2017). Therefore, all activities related to standards and regulation need to be carried out iteratively or in parallel and throughout the entire NPD process. Similarly, we also identify testing as a continuous activity. Starting testing early on and continuing it throughout the NPD process prevents potentially expensive re-work to change designs that do not conform to standards at a late stage in the process. Our study therefore highlights the need for an iterative approach in order to reap the benefits of standards outlined above.

---

[2] As we observed in our case when companies were initially unaware of important aspects of electricity generation where standards applied (see Sect. 4.2.1).

7.2.3   Addressing External Developments on the Industry Level and in the Wider Context

One of Adolphi's (1997) key findings relates to companies facing a 'make-or-buy decision' when they require standards. Our study clearly shows that innovating firms frequently face a similar choice between adapting their technology to standards and regulation or (attempting to) adapt(ing) standards and regulation to the technology. This choice applies in particular when addressing uncertainties resulting from a lack of needed standards (see Sect. 7.1.3). While this choice—to our knowledge—has not yet been documented in the standardisation literature, it closely resembles some strategies identified in studies on regulatory uncertainty (e.g. Engau & Hoffmann, 2011a, 2011b; Fremeth & Richter, 2011).

Such attempts to influence standards and regulation are the core channel through which companies can affect the dynamics on the industry level and in the technology's wider context. In line with earlier findings (e.g. de Vries et al., 2009; Foukaki, 2017; Jakobs, 2017), we show that this option is only open to companies with sufficient awareness of the topic, financial resources, expertise, etc. (see the argument above). This means that companies without these supporting factors have a very limited impact (if any at all) on external developments. De Vries et al. (2009) argue that they can be represented by trade associations (as we observed to some degree in our case). However, relying on such proxies implies (1) that this element of the industry structure (see Sect. 6.2.1) is sufficiently developed and (2) that industry associations act in line with the interests of member companies that do not engage in standardisation. Even when there are strong industry associations, the second assumption may not always be true: Our case shows associations are likely to be dominated by the same companies that are active in standardisation, because engaging in them is similarly resource intensive as participating in standardisation. Companies that engage neither in standardisation nor industry associations are therefore often 'standard takers' rather than 'standard makers' (see the distinction by Meyer, 2012) and interactions between the company level and external developments are mostly inwards-flowing for them through the activities discussed above.

Furthermore, companies that engage in standardisation and regulation need a long-term outlook. This is not only needed because standardisation and regulation processes tend to be lengthy, but also because of the 'public good nature' of standards (see Berg, 1989; Blind, 2006; Tassey, 2000). Standard takers eventually also enjoy many of the benefits from being able to access the market once standards and regulation have been adapted to the technology, but incur none of the costs. Standard makers need to accept that many (but not all) benefits of their work are public. Our study shows that they tend to be motivated by the opportunity to shape the contents of standards and regulation based on their individual preferences. In addition, the required standards and regulation are unlikely to be developed if no company takes action and everyone waits for other players to take the initiative.

Even if companies participate in standardisation and attempt to influence e regulation, they are unlikely to succeed in doing so on their own. Cooperation with others is therefore needed. A fundamental decision in this context revolved around which forums for collaboration to engage in. In this context, they need to navigate potentially complex interdependent arrangements of organisations, including SDOs, industry trade associations, and consortia, that might span across multiple modes of standardisation (see Wiegmann et al., 2017). While the motivations identified in the earlier literature for participating in these settings (Blind & Mangelsdorf, 2016; Jakobs, 2017) are confirmed by our study, it appears that different forums for cooperation may fulfil distinct functions in companies' strategies. For example, we observe an emphasis on technological knowledge sharing when participating in technology development consortia. In contrast, firms' activities in SDOs and industry associations appear to be more geared towards ensuring conformity to regulatory requirements and arranging compatibility with other elements of a large system in our case. Ultimately, all of these activities observed in our study were driven by the goal of building a market in which the technology could succeed. This market required rules in the form of standards (also see Fligstein & McAdam, 2012; Polanyi, 2001) as well as a critical mass for the technology.

Cooperation in technology development and pursuing changes to standards and regulation is one side of firms' engagement on the industry level and in the wider context. On the other side, they remain rivals and compete with each other once their products enter the market. Participating in the processes at the industry level and beyond therefore requires firms to follow a co-opetitive approach (see, e.g. Bengtsson & Kock, 2000; Gnyawali & Park, 2009, 2011; Van de Ven, 2005; Walley,

2007). We explore the dynamics that occur in such co-opetitive relationships in Sect. 7.3.

7.3 Dynamics on the Industry Level and Beyond

While the needed well-functioning system of standards (see Sect. 7.1.1) may often be taken for granted, it actually is the result of a very dynamic process. We expected in our literature review that this process would mainly take place at the industry level (see Sects. 1.2.2 and 1.2.3). Unexpectedly, our study revealed that the industry's wider context (which covers stakeholders outside the industry where the innovation is developed) also plays a very important role. This reflects research approaches which highlight the embedding of markets in society (Fligstein & McAdam, 2012; Polanyi, 2001). Addressing influences coming from this wider context is facilitated by strong cooperation among stakeholders in support of the innovation, both within the industry and across its boundaries.

Our study contributes to the literature on these dynamics in three ways: (1) We show what causes these dynamics (Sect. 7.3.1). (2) We then reveal industry-level approaches to address these dynamics (Sect. 7.3.2). (3) Following on from this, we argue that these dynamics allow standards to fulfil their function of aligning the innovation with the needs of the wider context (Sect. 7.3.3).

7.3.1 Sources of Dynamics in the Industry and Wider Context

Much of the dynamics in the process of establishing standards and regulation for an innovation are caused by conflicting interests of involved stakeholders. In our case, the aims of parties involved in developing the technologies were aligned, but even an innovation's developers do not always agree on a common direction. For example, strong differences could be observed among the developers of GSM (e.g. Bekkers, 2001) or in the case of e-mobility charging (Bakker, Leguijt, & van Lente, 2015; Wiegmann, 2013). Our study shows that this picture is further complicated by stakeholders who are not involved in developing the technology but are nevertheless affected by it. The types of interests pursued by these stakeholders can be very diverse and relate to many topics, such as preserving a status-quo that works for them, facilitating another technology that emerges in parallel, or government achieving its policy objectives.

This wide variety of interests and stakeholders, which can potentially be affected by the standardisation and regulation of an innovation, causes the core of the dynamics in the process. All involved parties can potentially intervene in the process at any time (see Wiegmann et al., 2017), either to support the innovation or to hinder it. In that context, we observed many different tactics to reach these goals. This wide range of tactics includes attempts to use standards as a tool to actively block a technology (also see Delaney, 2001), coalition building (also see Axelrod et al., 1995), or lobbying the government to intervene (also see Wiegmann et al., 2017). This potential variety of tactics also causes challenges for managing standards and regulation on the industry level, as we outline below.

7.3.2   Industry-Level Approaches for Addressing Dynamics in the Process

The dynamics discussed above challenge the view taken by some that the development of standards to support an innovation can be planned and coordinated by a central actor, such as a government (Featherston et al., 2016; Ho & O'Sullivan, 2017). Although governments (or other actors) sometimes play such a central role, others still can use a range of channels to challenge this (this study; Wiegmann et al., 2017). It may be possible to forecast at what stage of a technology trajectory certain standards would be needed through roadmapping and other tools (Blind & Gauch, 2009; Featherston et al., 2016; Ho & O'Sullivan, 2017). However, the actual emergence of such standards depends on whether the involved parties reach a balance of interests and whether they can sustain this compromise.

Nevertheless, our study shows that there are a number of ways to facilitate this outcome, if not to plan it. Strong collaboration among a technology's supporters and with industry-external actors who share the same or complementary interests is at the core of this. Our study highlights several factors that can support such cooperation and help the industry as a whole to navigate the dynamics in a way that increases the likelihood of establishing standards and regulation which support an innovation. Below, we discuss the role of supporting institutions and an optimal approach to IPR as factors that stand out as particularly important for this collaboration. Following this, we address our findings regarding the resulting 'group dynamics'.

7.3.2.1 Supporting Institutions for Effective Collaboration

A first core element of our findings is the importance of an industry's supporting institutions, e.g. industry associations. They can enhance cooperation in a number of ways, e.g. by providing forums in which actors can agree on common positions to pursue (similar to the role of consortia observed by Baron et al. (2014) in ICT standardisation), or by implementing common technology development initiatives. In addition to facilitating industry-internal alliances, such supporting institutions may also have established links to actors in the wider context (e.g. governments, trade associations in other industries) that can be used strategically to influence standards and regulation in the technology's favour.

7.3.2.2 The Importance of Intellectual Property Rights in Effective Collaboration

A second factor underlying effective collaboration is an appropriate approach to IPR. Here, our study questions whether the widely held view of a tight link between standards and patents (e.g. Bekkers, 2017; Bekkers, Iversen, & Blind, 2011; Großmann et al., 2016; Lerner & Tirole, 2014; Rysman & Simcoe, 2008) always applies. Patents have been identified as a core element of many standardisation processes. However, giving them a similar role in our case would have undermined both effective collaboration within the industry, and the degree to which the resulting standards would have been perceived legitimate by others. Indeed, the involved parties aimed to keep patents as separate from standards as possible, although they still gave them a prominent role in the collaborations to develop the technology. The industry in our study managed to find a fine balance between protecting firms' intellectual input into the technology's development, while not crowding others out of the process.

To understand these different findings, we contrast our case to others where intellectual property played a more important role, such as mobile telecommunications (see, e.g. Bekkers, 2001; Funk & Methe, 2001; Leiponen, 2008), Ethernet (see Jain, 2012; von Burg, 2001), and optical disks (see den Uijl, Bekkers, & de Vries, 2013). This suggests that the type of standards that are being developed is core to the importance of patents in the process: Many cases where patents were important concern interface standards (see the classifications by Blind, 2004, 2017; Egyedi & Ortt, 2017; Swann, 2010), which are by definition solution-prescribing (see, e.g. de Vries, 1998; Tassey, 2000). Such solutions are based on concrete designs that are usually patentable. On the other hand, most standards in our case fulfilled economic functions related to safety and measurement and were performance-based, meaning that little (if any) of their content could be patented.

However, not all standards in our case were performance-based: For example, standards for connecting to the electricity grid had important interface elements and therefore incorporated patentable solutions. Nevertheless, we also did not observe an important role of IPR in these standards' development. This can be explained by the 'standardisation culture' that applies in a specific context (see Wiegmann et al., 2017). In the industries in our case, this 'culture' clearly is collaborative and long-term oriented, and most standards that we found link strongly to regulation. This would make any attempts of bringing patents into standardisation unacceptable to many stakeholders. In other industries, such as ICT, most standards arguably concern interfaces that are based on the private intellectual property, and have few links to regulation. Under such circumstances, it is no surprise that the common approach to standardisation emphasises patents more.

In summary, the different emphasis on patents in standardisation is initially likely to result from the types of standards that prevail in an industry. This emphasis is then likely to perpetuate itself and become a part of the industries 'standardisation culture'.

7.3.2.3 'Group Dynamics' Resulting from the Collaboration in an Industry

The activities (both in terms of technology development and standardisation/regulation), which make up the cooperation in the industry, contribute to certain 'group dynamics'. In our case, we observed a strongly united industry with an 'us vs. them' mentality in its relations to other stakeholders. In other cases, these group dynamics may vary depending on the distribution of interests and contextual factors like the 'standardisation culture' (see Wiegmann et al., 2017). Our study suggests that such group dynamics affect the degree to which the innovators' activities are perceived as legitimate (see Botzem & Dobusch, 2012; Delemarle, 2017; Tamm Hallström & Boström, 2010) by other actors in the wider context. In particular, Botzem and Dobusch's (2012) concept of standards' input legitimacy is likely to be strongly affected by the composition of an innovation's group of supporters and their activities. For example, in our case, the industry speaking with one voice signalled that mCHP was a genuine technological development for which changing standards and regulation was warranted, rather than a single company's attempt to get special treatment. However, this approach also carried the danger of being perceived as an industry that writes its own rules, similar to the European car industry in the wake of the Volkswagen Diesel scandal (see Neslen, 2015). Our study therefore suggests that the collaborative activities of an innovation's supporters have an important impact on the perceived legitimacy. Future research could compare different approaches and their effects in this regard, e.g. by involving more stakeholders (see Sect. 7.5).

7.3.3 Dynamics' Support for Aligning the Innovation with the Wider Context

In Sects. 6.4 and 7.1.1, we argued that standards fulfil an important function in aligning the innovation with the needs of relevant stakeholders in the technology's wider context. Arguably, the dynamics discussed in this chapter are core to standards fulfilling this function, because they end in the balance that stakeholders must reach for a standard to emerge (see Wiegmann et al., 2017). In that sense, the dynamic processes in standardisation and regulation that we observed are an important element of the wider sociotechnical transition needed to make an innovation successful. In such sociotechnical transitions, innovations either move out of the niches in which they emerge by reaching alignment with the sociotechnical system that are part of, or they fail eventually (e.g. Geels & Schot, 2007; Smith & Raven, 2012; Smith et al., 2010; van den Ende & Kemp, 1999).

By specifying clear technological requirements that result from the needs of other actors in the sociotechnical environment and the sociotechnical system (in our case, e.g. related to CO_2 emission targets, or the needs of other users of the electricity grid for grid stability), standards and regulation contribute to this alignment. This function explains the high stakes at play that lead to the dynamics that we observed. Simultaneously, we argue that standards would not be able to fulfil this function in support of sociotechnical transitions without these dynamics. A less dynamic process could most likely only be achieved if it failed to take into account some of the diverse interests typically involved in sociotechnical transitions. The resulting standards would therefore not align the innovation with the needs of its wider context and miss important benefits for the innovation outlined in Sect. 7.1.

7.4   Managerial Implications

Our findings also have strong implications for managerial practice. In particular, we offer insights on three topics that are highly relevant for innovative companies: (1) We highlight important effects of standards (Sect. 7.4.1). (2) We show how innovators can successfully address standards and regulation (Sect. 7.4.2). (3) We identify impactful dynamics on the industry level and beyond, and show how they can be managed through cross-company collaboration (Sect. 7.4.3).

7.4.1   Important Effects of Standards

Standards can have major positive effects on innovation, such as supporting the technology's legitimacy, securing the links between complementary products, and facilitating proof of regulatory compliance. On the other hand, standards which are not in line with an innovation's needs can impose substantial hurdles, e.g. if standards lock the market into an old technology, or reflect vested interests that oppose the innovation. However, we find no support for the popular assumption that standards in general limit the freedom of innovation. Instead, the freedom for innovating depends on how well standards are managed and integrated in the innovation process (see Sects. 6.1.3 and 7.2).

In the European context, standards often are linked to regulation. This link further amplifies their effects on innovation. Harmonised standards, which are in line with an innovation's needs, can be used to show regulatory compliance and give innovators a high degree of legal certainty. On the other hand, innovators can face substantial costs and difficulties in proving regulatory compliance if harmonised standards are not in line with their innovation's needs. The required effort may sometimes even be prohibitively high, meaning that such standards can effectively lock an innovation out of the market.

The possible magnitude of standards' effects makes them a topic that innovation managers need to be aware of. Furthermore, they also mean that missing standards are an important factor causing uncertainty when innovating. Fortunately, an innovation's developers can actively manage standards and their effects. Our study provides managers with useful insights into how this can be done effectively, as we outline in Chapter 6, Sects. 7.2 and 7.3.

7.4.2 Implications for Company-Internal Management

Our study shows successful approaches that companies can use to manage the effects of standards on their innovations. Within these approaches, we distinguish between the organisational foundation and the specific management activities.

In the long term, companies need to prepare themselves for dealing with standards and regulation. To do so, they should establish a solid organisational foundation that allows them to take a strategic approach to standards and regulation. Such a foundation is rooted in awareness, expertise, and financial resources. For large companies, this may mean establishing a department that is responsible for coordinating the topic. Small companies should aim to have at least some staff members with awareness and basic knowledge of standardisation and regulation. Such internally developed competences can be complemented by external experts (e.g. consultants, notified bodies). However, our study shows that relying on them too heavily may limit the company's freedom in innovating.

Such a foundation helps companies to carry out the activities needed to manage the topic: (1) identifying regulation and standards, (2) specifying the product, (3) assessing whether modifications in standards/regulation and/or the product design are needed, and, if necessary, (4) engaging in standardisation. Because firms operate in a dynamic environment, these activities need to be carried out concurrently and throughout the NPD process. This means that companies should identify potentially relevant regulation and standards as early as possible and then continue scanning for potential changes or additional requirements that they missed at first. It also means that the NPD process should involve regular checks whether the design is capable of meeting all requirements. Doing so in parallel avoids both being blindsided by changes in standards and regulation and having to redo large parts of the innovation if certain requirements cannot be met.

A further key decision is whether companies limit themselves to applying standards and regulation to their innovations or whether they also attempt to influence standardisation and the passing of new regulation. Companies that do not engage in such external activities still benefit from the results of others that do. However, our findings suggest that this engagement has benefits, which often may justify the necessary expenditure. Most importantly, companies that contribute to external standardisation and regulation processes have an opportunity to participate in shaping the balance of interests enshrined in standards in their favour (see Sect. 7.3). This may substantially increase the company's freedom innovating.

7.4.3   Implications for Cross-Company Collaboration

Our study shows that these company-external processes are likely to be highly dynamic. These dynamics result from a potentially large number of stakeholders with conflicting interests, all of whom are likely to attempt influencing standards and regulation in their favour. Our study shows that even innovations like mCHP, which are relatively simple and small innovations,[3] can have substantial links to the wider context and affect many parties' interests. In addition to stakeholders from innovators' own industries, these stakeholders therefore often include actors from the wider context (e.g. regulators, developers of other technologies, NGOs).

Few companies (if any) are likely to be strong enough to be able to shift standards on their own under these conditions. Cooperation in developing both the technology and relevant standards is therefore at the core of influencing external standardisation and regulation. Consequently, innovative companies need to find partners who can complement their own strengths. This cooperation fulfils multiple functions, such as aligning industry actors to pursue a common line in standardisation, and legitimising the technology in the eyes of outsiders.

Reaching these goals can be supported by an industry structure that enables effective collaboration. We identify three elements of the industry structure that are important in this context: (1) a network of supporting institutions (e.g. industry associations, consultants, research institutions), (2) an approach to IPR that facilitates cooperation, and (3) broad support for the innovation among firms in the industry. These

---

[3] Compared to large-scale systems like autonomous driving and Smart Cities.

three elements can support collaboration in many ways. For example, they can help resolve conflicts (or even prevent them from occurring), unlock additional sources of helpful expertise, and provide access to regulators. Companies and other actors in an industry are therefore advised to build these elements in time, so that they are available when needed.

We also show that basing industry-level collaboration on this support structure helps innovators to assert themselves in dealing with the complex dynamics of their industry's wider context, as the following three examples show. (1) Industry associations can help unite the industry behind an innovation, giving it a stronger voice when dealing with other stakeholders. (2) Involving other supporting actors, who have no direct commercial interest in the technology (e.g. researchers), can help the innovation's legitimacy and credibility. (3) Using suitable approaches to IPR in standardisation may make it more acceptable to link the resulting standards to regulation.

This also makes our findings important for actors other than companies. Especially industry associations can assume an important role in coordinating the collaboration between their members. For example, they can offer forums for industry to find a common position to pursue in standardisation committees and vis-à-vis regulators. They can also represent industry when dealing with external stakeholders on aspects that are not central to the innovation, but nevertheless need to be considered.

7.5   Limitations and Scope for Further Research

Our detailed grounded theory study provides novel insights into the management of standards as an example of the external requirements, which innovative companies face. First, this raises the question under which conditions our theory is likely to apply (Sect. 7.5.1). Furthermore, the results raise intriguing questions for future research (Sect. 7.5.2).

7.5.1   Generalising Our Theory

Our theory is based on a single nested case. This means that the company-level findings have undergone an initial replication (see Eisenhardt, 1989; Yin, 2009) whereas the industry-level elements of our theory are derived from a single observation. Nevertheless, we expect that similar observations can be made in other cases which share several key characteristics, which likely determined parts of what we witnessed with our case. These key features of the case are (1) its European scope (due to the relationships of standards and regulation under the 'New Approach'); (2) the highly regulated nature of the industry on aspects like product safety which contributed to the particular importance of standards in the case; (3) the relationship with policy issues (energy and environmental policy in our case); and (4) the relative long-term outlook of the key players in the case which contributes to the industry's culture of collaboration. Other areas where we expect that cases with similar characteristics to exist include, e.g. the European medical and aerospace sectors. In addition to the factors outlined above, the 'self-evident' support for standards in our case most likely makes it a 'best practice case'. Future research therefore needs to confirm the extent to which our findings apply to both similar and other contexts, which do not share the four characteristics identified above. It also needs to establish the extent to which not following the practices identified in our case affects innovation.

7.5.2   Questions for Future Research

Many of our study's new insights raise questions that could lead to exciting new research. Some of them question findings in previous standardisation literature, whereas others point to links with other streams of literature that have not yet been explored extensively.

One issue that raises questions for future research is IPR's relatively low importance for standardisation in the heating sector (see Sect. 5.1.4). This raises doubts about the standardisation literature's emphasis on IPR. This emphasis may be related to the literature's empirical evidence largely coming from the ICT sector (see Wiegmann et al., 2017). Future research in other settings could establish whether our case is an anomaly and IPR is indeed as important for standardisation as the literature claims, or whether this only applies to ICT contexts. In doing so, such research should also consider factors like the type of standard at stake and the 'standardisation culture' that we identify as potentially important for the role of IPR in standardisation (see Sect. 7.3.2).

The most intriguing questions for future research relate to the link between standardisation and the wider context. Previous literature on the co-evolution between standards and innovation (e.g. Blind & Gauch, 2009; Featherston et al., 2016; Ho & O'Sullivan, 2017) does not emphasise this link and mostly focuses on the industry. Consequently, the significance of this link was a surprising finding, which we did not anticipate when planning our study. Our theory identifies two important patterns related to this link (diverse types of interests and strategies for dealing with them, see Sect. 6.3), which were consistently addressed across interviews. However, in line with our research question's focus on innovators' management, we did not interview actors in the wider context. This means that more than the two prominent patterns, which we already identify, may exist in this link, e.g. related to impacts on large societal trends. Future in-depth research, which builds on this contribution, is needed therefore to completely uncover the connection between innovation, standards, and the wider context.

This research would potentially contribute to streams of literature beyond standardisation: Related to sociotechnical systems theories (e.g. Geels, 2004; Geels & Schot, 2007; Smith & Raven, 2012; Smith et al., 2010), the research could potentially offer new insights into how transitions occur and how they are supported by standards. In that context, research on the link between standards and the wider context could also contribute to theories on the needs of rules underlying markets (e.g. Fligstein & McAdam, 2012; Polanyi, 2001) and on regulatory uncertainty (e.g. Engau & Hoffmann, 2011a, 2011b).

Potentially, such research could build on the emerging literature that links co-opetition to standards (e.g. Allamano-Kessler, Mione, & Larroque, 2016; Benmeziane & Mione, 2016; Foukaki, 2017). As we argue in Sect. 7.3, co-opetitive approaches are likely to have a substantial effect on how the legitimacy of both an innovation and the applicable regulation and standards are perceived by stakeholders in the wider context. Future research could take this finding as a basis, for example to identify whether specific co-opetition patterns are particularly conducive to building legitimacy.

References

Abraham, J., & Reed, T. (2002). Progress, innovation and regulatory science in drug-development: The politics of international standard-setting. *Social Studies of Science, 32*(3), 337–369. https://doi.org/10.1177/03063127020 32003001.

Adolphi, H. (1997). *Strategische Konzepte zur Organisation der betrieblichen Standardisierung*. Berlin, Vienna, Zürich: Beuth Verlag.

Allamano-Kessler, R., Mione, A., & Larroque, L. (2016). Fatal competition, peaceful coexistence or active coopetition between traceability standards in the distribution channel? In K. Jakobs, A. Mione, A.-F. Cutting-Decelle, & S. Mignon (Eds.), *EURAS proceedings 2016—Co-opetition and open innovation* (pp. 1–18). Aachen: Verlagshaus Mainz.

Allen, R. H., & Sriram, R. D. (2000). The role of standards in innovation. *Technological Forecasting and Social Change, 64*(2–3), 171–181. https://doi.org/10.1016/S0040-1625(99)00104-3.

Axelrod, R., Mitchell, W., Thomas, R. E., Bennett, D. S., & Bruderer, E. (1995). Coalition formation in standard-setting alliances. *Management Science, 41*(9), 1493–1508.

Bakker, S., Leguijt, P., & van Lente, H. (2015). Niche accumulation and standardization—The case of electric vehicle recharging plugs. *Journal of Cleaner Production, 94,* 155–164. https://doi.org/10.1016/j.jclepro.2015.01.069.

Baron, J., Ménière, Y., & Pohlmann, T. (2014). Standards, consortia, and innovation. *International Journal of Industrial Organization, 36*(9), 22–35. https://doi.org/10.1016/j.ijindorg.2014.05.004.

Bekkers, R. (2001). *Mobile telecommunications standards: UMTS, GSM, TETRA, and ERMES.* Boston, MA: Artech House.

Bekkers, R. (2017). Where patents and standards come together. In R. Hawkins, K. Blind, & R. Page (Eds.), *Handbook of innovation and standards* (pp. 227–251). Cheltenham: Edward Elgar.

Bekkers, R., Iversen, E., & Blind, K. (2011). Emerging ways to address the reemerging conflict between patenting and technological standardization. *Industrial and Corporate Change, 21*(4), 901–931. https://doi.org/10.1093/icc/dtr067.

Bengtsson, M., & Kock, S. (2000). "Coopetition" in business networks—To cooperate and compete simultaneously. *Industrial Marketing Management, 29,* 411–426.

Benmeziane, K., & Mione, A. (2016). Coopetition to gain influence, leadership and control on standard setting organization. In K. Jakobs, A. Mione, A.-F. Cutting-Decelle, & S. Mignon (Eds.), *EURAS proceedings 2016—Co-opetition and open innovation.* Aachen: Verlagshaus Mainz.

Berg, S. V. (1989). Technical standards as public goods: Demand incentives for cooperative behavior. *Public Finance Review, 17*(1), 29–54. https://doi.org/10.1177/109114218901700102.

Blind, K. (2004). *The economics of standards—Theory, evidence, policy.* Cheltenham: Edward Elgar.

Blind, K. (2006). Explanatory factors for participation in formal standardisation processes: Empirical evidence at firm level. *Economics of Innovation and New Technology, 15*(2), 157–170. https://doi.org/10.1080/10438590500143970.

Blind, K. (2017). *The economic functions of standards in the innovation process.* In R. Hawkins, K. Blind, & R. Page (Eds.), *Handbook of innovation and standards* (pp. 38–62). Cheltenham: Edward Elgar. http://doi.org/10.4337/9781783470082.

Blind, K., & Gauch, S. (2009). Research and standardisation in nanotechnology: Evidence from Germany. *The Journal of Technology Transfer, 34*(3), 320–342. http://doi.org/10.1007/s10961-008-9089-8.

Blind, K., & Mangelsdorf, A. (2016). Motives to standardize: Empirical evidence from Germany. *Technovation, 48–49,* 13–24. https://doi.org/10.1016/j.technovation.2016.01.001.

Borraz, O. (2007). Governing standards: The rise of standardization processes in France and in the EU. *Governance, 20*(1), 57–84. https://doi.org/10.1111/j.1468-0491.2007.00344.x.

Botzem, S., & Dobusch, L. (2012). Standardization cycles: A process perspective on the formation and diffusion of transnational standards. *Organization Studies, 33*(5–6), 737–762. https://doi.org/10.1177/0170840612443626.

David, P. A. (1985). Clio and the economics of QWERTY. *The American Economic Review, 75*(2), 332–337.

de Vries, H. J. (1998). The classification of standards. *Knowledge Organization, 25*(3), 79–89.

de Vries, H. J. (1999). *Standardization—A business approach to the role of national standardization organizations.* Boston, Dordrecht, and London: Kluwer Academic Publishers.

de Vries, H. J., Blind, K., Mangelsdorf, A., Verheul, H., & van der Zwan, J. (2009). *SME access to European standardization: Enabling small and medium-sized enterprises to achieve greater benefit from standards and from involvement in standardization.* Rotterdam:(give space) Rotterdam School of Management, Erasmus University.

de Vries, H. J., & Verhagen, W. P. (2016). Impact of changes in regulatory performance standards on innovation: A case of energy performance standards for newly-built houses. *Technovation, 48–49,* 56–68. https://doi.org/10.1016/j.technovation.2016.01.008.

Delaney, H. (2001). Standardization and technical trade barriers: A case in europe. In S. M. Spivak & F. C. Brenner (Eds.), *Standardization essentials: principles and practice* (pp. 161–166). New York, NY: Marcel Dekker.

Delemarle, A. (2017). Standardization and market framing: The case of nanotechnology. In R. Hawkins, K. Blind, & R. Page (Eds.), *Handbook of innovation and standards* (pp. 353–373). Cheltenham: Edward Elgar. http://doi.org/10.4337/9781783470082.

den Uijl, S., Bekkers, R., & de Vries, H. J. (2013). Managing intellectual property using patent pools: Lessons from three generations of pools in the optical disc industry. *California Management Review, 55*(4), 31–50. https://doi.org/10.1525/cmr.2013.55.4.31.

Egyedi, T. M., & Heijnen, P. (2008). How stable are IT standards? In T. M. Egyedi & K. Blind (Eds.), *The dynamics of standards* (pp. 137–154). Cheltenham: Edward Elgar.

Egyedi, T. M., & Ortt, J. R. (2017). Towards a functional classification of standards for innovation research. In R. Hawkins, K. Blind, & R. Page (Eds.), *Handbook of innovation and standards* (pp. 105–134). Cheltenham: Edward Elgar. http://doi.org/10.4337/9781783470082.

Eisenhardt, K. M. (1989). Building theories from case study research. *Academy of Management Review, 14*(4), 532–550. https://doi.org/10.5465/AMR.1989.4308385.

Engau, C., & Hoffmann, V. H. (2011a). Corporate response strategies to regulatory uncertainty: Evidence from uncertainty about post-Kyoto regulation. *Policy Sciences, 44*(1), 53–80. https://doi.org/10.1007/s11077-010-9116-0.

Engau, C., & Hoffmann, V. H. (2011b). Strategizing in an unpredictable climate: Exploring corporate strategies to cope with regulatory uncertainty. *Long Range Planning, 44*(1), 42–63. https://doi.org/10.1016/j.lrp.2010.11.003.

Featherston, C. R., Ho, J.-Y., Brévignon-Dodin, L., & O'Sullivan, E. (2016). Mediating and catalysing innovation: A framework for anticipating the standardisation needs of emerging technologies. *Technovation, 48–49,* 25–40. https://doi.org/10.1016/j.technovation.2015.11.003.

Fligstein, N., & McAdam, D. (2012). *A theory of fields.* New York: Oxford University Press.

Foukaki, A. (2017). *Corporate standardization management: A case study of the automotive industry.* Lund: Lund University. Retrieved from http://portal.research.lu.se/ws/files/21522119/Corporate_Standardization_Management_A_Case_Study_of_the_Automotive_Industry_Dissertation_2017.pdf.

Fremeth, A. R., & Richter, B. K. (2011). Profiting from environmental regulatory uncertainty: Integrated strategies for competitive advantage. *California Management Review, 54*(1), 145–165. https://doi.org/10.1525/cmr.2011.54.1.145.

Funk, J. L., & Methe, D. T. (2001). Market- and committee-based mechanisms in the creation and diffusion of global industry standards: The case of mobile communication. *Research Policy, 30*(4), 589–610. https://doi.org/10.1016/S0048-7333(00)00095-0.

Geels, F. W. (2004). From sectoral systems of innovation to socio-technical systems. *Research Policy, 33*(6–7), 897–920. https://doi.org/10.1016/j.respol.2004.01.015.

Geels, F. W., & Schot, J. (2007). Typology of sociotechnical transition pathways. *Research Policy, 36*(3), 399–417. https://doi.org/10.1016/j.respol.2007.01.003.

Gnyawali, D. R., & Park, B. R. (2009). Co-opetition and technological innovation in small and medium-sized enterprises: A multilevel conceptual model. *Journal of Small Business Management, 47*(3), 308–330. https://doi.org/10.1111/j.1540-627X.2009.00273.x.

Gnyawali, D. R., & Park, B. R. (2011). Co-opetition between giants: Collaboration with competitors for technological innovation. *Research Policy, 40*(5), 650–663. https://doi.org/10.1016/j.respol.2011.01.009.

Großmann, A.-M., Filipović, E., & Lazina, L. (2016). The strategic use of patents and standards for new product development knowledge transfer. *R&D Management, 46*(2), 312–325. https://doi.org/10.1111/radm.12193.

Ho, J., & O'Sullivan, E. (2017). Strategic standardisation of smart systems: A roadmapping process in support of innovation. *Technological Forecasting and Social Change, 115,* 301–312. https://doi.org/10.1016/j.techfore.2016.04.014.

Hoffmann, V. H., Trautmann, T., & Schneider, M. (2008). A taxonomy for regulatory uncertainty—Application to the European Emission Trading Scheme. *Environmental Science & Policy, 11*(8), 712–722. https://doi.org/10.1016/j.envsci.2008.07.001.

Jain, S. (2012). Pragmatic agency in technology standards setting: The case of Ethernet. *Research Policy, 41*(9), 1643–1654. https://doi.org/10.1016/j.respol.2012.03.025.

Jakobs, K. (2017). Corporate standardization management and innovation. In R. Hawkins, K. Blind, & R. Page (Eds.), *Handbook of innovation and standards* (pp. 377–397). Cheltenham: Edward Elgar. http://doi.org/10.4337/9781783470082.

Kondo, Y. (2000). Innovation versus standardization. *The TQM Magazine, 12*(1), 6–10. Retrieved from http://www.emeraldinsight.com/journals.htm?articleid=841925&show=abstract.

Leiponen, A. E. (2008). Competing through cooperation: The organization of standard setting in wireless telecommunications. *Management Science, 54*(11), 1904–1919. https://doi.org/10.1287/mnsc.1080.0912.

Lerner, J., & Tirole, J. (2014). A better route to tech standards. *Science, 343*(6174), 972–973. https://doi.org/10.1126/science.1246439.

Lorenz, A., Raven, M., & Blind, K. (2017). The role of standardization at the interface of product and process development in biotechnology. *The Journal of Technology Transfer,* 1–37. http://doi.org/10.1007/s10961-017-9644-2.

Meyer, N. (2012). *Public intervention in private rule making: The role of the european commission in industry standardization.* The London School of Economics and Political Science (LSE). Retrieved from http://etheses.lse.ac.uk/236/.

Neslen, A. (2015). *EU caves into auto industry pressure for weak emissions limits.* Retrieved August 31, 2016, from http://www.theguardian.com/environment/2015/oct/28/eu-emissions-limits-nox-car-manufacturers.

Polanyi, K. (2001). *The great transformation—The political and economic origins of our time* (2nd Beacon paperback). Boston, MA: Beacon Press.

Rysman, M., & Simcoe, T. (2008). Patents and the performance of voluntary standard-setting organizations. *Management Science, 54*(11), 1920–1934. https://doi.org/10.1287/mnsc.1080.0919.

Smith, A., & Raven, R. (2012). What is protective space? Reconsidering niches in transitions to sustainability. *Research Policy, 41*(6), 1025–1036. https://doi.org/10.1016/j.respol.2011.12.012.

Smith, A., Voß, J.-P., & Grin, J. (2010). Innovation studies and sustainability transitions: The allure of the multi-level perspective and its challenges. *Research Policy, 39*(4), 435–448. https://doi.org/10.1016/j.respol.2010.01.023.

Swann, G. M. P. (2010). *The economics of standardization: An update*. Retrieved March 21, 2013, from http://www.bis.gov.uk/assets/biscore/innovation/docs/e/10-1135-economics-of-standardization-update.pdf.

Tamm Hallström, K., & Boström, M. (2010). *Transnational multi-stakeholder standardization: Organizing fragile non-state authority*. Cheltenham: Edward Elgar.

Tassey, G. (2000). Standardization in technology-based markets. *Research Policy, 29*(4–5), 587–602. https://doi.org/10.1016/S0048-7333(99)00091-8.

Teece, D. J. (1986). Profiting from technological innovation: Implications for integration, collaboration, licensing and public policy. *Research Policy, 15*(6), 285–305. https://doi.org/10.1016/0048-7333(86)90027-2.

Teece, D. J. (2006). Reflections on "profiting from innovation". *Research Policy, 35*(8), 1131–1146. https://doi.org/10.1016/j.respol.2006.09.009.

Van de Ven, A. H. (1993). A community perspective on the emergence of innovations. *Journal of Engineering and Technology Management, 10*(1–2), 23–51. https://doi.org/10.1016/0923-4748(93)90057-P.

Van de Ven, A. H. (2005). Running in packs to develop knowledge-intensive technologies. *MIS Quarterly, 29*(2), 365–378. https://doi.org/10.2307/25148683.

van den Ende, J., & Kemp, R. (1999). Technological transformations in history: How the computer regime grew out of existing computing regimes. *Research Policy, 28*(8), 833–851. https://doi.org/10.1016/S0048-7333(99)00027-X.

van Schewick, B. (2010). *Internet architecture and innovation*. Cambridge, MA; London: The MIT Press.

van Wessel, R. (2010). *Toward corporate it standardization management—Frameworks and solutions*. Hershey, PA: Information Science Reference.

von Burg, U. (2001). *The triumph of Ethernet: Technological communities and the battle for the LAN standard*. Stanford: Stanford University Press.

Wakke, P., Blind, K., & De Vries, H. J. (2015). Driving factors for service providers to participate in standardization: Insights from the Netherlands.

Industry and Innovation, 22(4), 299–320. https://doi.org/10.1080/13662 716.2015.1049865.

Walley, K. (2007). Coopetition: An introduction to the subject and an agenda for research. *International Studies of Management and Organization, 37*(2), 11–31. https://doi.org/10.2753/IMO0020-8825370201.

Wiegmann, P. M. (2013). Combining different modes of standard setting—Analysing strategies and the case of connectors for charging electric vehicles in Europe. In K. Jakobs, H. J. de Vries, A. Ganesh, A. Gulacsi, & I. Soetert (Eds.), *EURAS proceedings 2013—Standards: Boosting European competitiveness* (pp. 397–411). Aachen: Wissenschaftsverlag Mainz.

Wiegmann, P. M., de Vries, H. J., & Blind, K. (2017). Multi-mode standardisation: A critical review and a research agenda. *Research Policy, 46*(8), 1370–1386. https://doi.org/10.1016/j.respol.2017.06.002.

Yin, R. K. (2009). *Case study research—Design and methods* (4th ed.). Thousand Oaks, CA: Sage.

INDEX

© The Editor(s) (if applicable) and The Author(s) 2019
P. M. Wiegmann, *Managing Innovation and Standards*,
https://doi.org/10.1007/978-3-030-01532-9